Pvt. Ltd.

Humic Substances in Drug Development

Dr. Rajesh Khanna
M.Pharm, Ph.D. (Jamia Hamdard)
Current affiliation: Dabur Pharma Ltd.,
Sahibabad, Ghaziabad, UP (INDIA)

•

Dr. S.P. Agarwal
B. Pharm. (Pilani), *M. Pharm.* (B.H.U.), *M.S., Ph.D.* (Illinois)
Department of Pharmaceutics
Faculty of Pharmacy
Jamia Hamdard (Hamdard University)
New Delhi (INDIA)

•

Dr. R.K. Khar
M. Pharm., D.B.M., Ph.D. (Sofia)
Department of Pharmaceutics
Faculty of Pharmacy
Jamia Hamdard (Hamdard University)
New Delhi (INDIA)

2009

Studium Press (India) Pvt. Ltd.

Humic Substances in Drug Development

ISBN: 978-81-907577-1-3

Published by:

Studium Press (India) Pvt. Ltd.
4735/22, 2nd Floor, Prakash Deep Building
(Near Delhi Medical Association),
Ansari Road, Darya Ganj, New Delhi-110 002
Tel.: 23240257, 65150447; Fax: 91-11-23240273;
jngovil@studiumpress.com; shray@studiumpress.com

Printed at:

Salasar Imaging Systems
Delhi-110035 (India)

FOREWORD

Knowledge has been passed on from generation to generation by various means. Cultural variations are common and affect knowledge transfer. Talking, listening, understanding, and writing (documenting) are essential steps in this process. Traditional knowledge of medicine has also followed either of these routes and has passed on from one generation to another. It is only recently that followers of modern system of medicine have started evaluating and understanding this traditional knowledge using modern analytical and pharmacological tools. The present book is also the result of one such evaluation and understanding, by the authors and other groups of scientists, of the nature and properties of Shilajit and similar humic substances which have been used the world-over as part of traditional medicine.

I vividly remember my meeting when the authors of this book, Rajesh Khanna and Suraj P. Agarwal met me some time in 2001 at my previous work place, for discussions on what to research from Ayurveda. They listened to my hypothesis on Shilajit and its components namely "humic acids and fulvic acids" (a product that Dabur Research Foundation was working for standardization) and my own belief that these form structures which look closely similar to Cyclodextrin. Shilajit is known to exhibit a 'bio-availability enhancer" property as one of its properties as per Ayurveda. Hence, I had felt that such an activity could be attributable to the presence of humic and fulvic acids, and a good scientific evaluation of same would be worth while. Such studies would not only validate the Ayurvedic knowledge, but may open up new vistas in drug development and designs.

It is heartening that the authors and their colleagues have taken this work so seriously and generated scientific data by working on various aspects of drug development using humic and fulvic acids. This well referenced book is the result of painstaking efforts for over 7 years. It is yet another example of an Ayurvedic ingredient or component of an Ayurvedic ingredient examined scientifically for its veracity and applications. They have adopted appropriate methods of science and technology and have multiple examples of usage of these substances for many pharmaceuticals. Inclusion of safety data adds huge value to the investigations. In addition, it must attract Pharmaceutical firms to take it up for drug development and bring marketable products, thus bringing glory to an ancient Ayurvedic knowledge used even today. Somehow many of us and even the country believes that acceptance of Ayurveda by the West by getting approval for marketing them as Drugs, in addition to them being marketed as Ayurvedic medicines or as supplements is the final word, and this book provides so much of information to do so. It is also another example of demonstration of "Reverse Pharmacology" successfully. Humic and fulvic acids ability to form complexes which benefit drug availability and its potential to improve therapeutic benefits creates a strong case for this natural substance as a useful pharmaceutical aid.

I am happy to have been asked to write a foreword to this book, and compliment the authors and the publishers for their successful effort. The book truly "exhumes humic and fulvic acids" and provides every thing that one would like to have at one place. More such champions are needed today, and hope that this book would enthuse researchers and entrepreneurs to champion the cause.

D.B. Anantha Narayana,
M.Pharm, Ph.D.,
Current Affiliation: Head, Herbals Research,
Unilever Research India,
64, Main Road, Whitefield,
Bangalore 560066.

PREFACE

Over the last century, several groups of scientists have worked in different regions of the world to study the pharmacological and medicinal nature of various groups of humic substances. However, the work has mainly been regional and did not find an international recognition. The main reasons for this could be attributed to:

- Diverse nature of humic substances per se.
- Lack of standardization of the natural material.
- Lack of complete characterization of the material and assignment of a molecular structure.
- Lack of clinical studies as per acceptable norms in modern system of medicine.
- Conflicting reports regarding the origin, nature, components and pharmacological properties.
- A number of synonyms and vernacular names for the same substance.
- The regional nature of the work carried out.
- Publication of the work in regional languages.

The present book is an attempt to collect the diverse information available on the medicinal aspects of humic substances, mainly shilajit in one cover. It is a single-source reference work to cover the classic as well as modern studies on this traditional herbo-mineral ancient remedy and includes scientific studies carried out by the authors demonstrating the drug complexing and bioavailability enhancing properties of its components which can advantageously be used for increasing the solubility, dissolution, permeability and hence bioavailability of problematic drugs of modern system of medicine.

This book aims to highlight the ancient claims of Shilajit, how these have been validated and certain properties of shilajit which can advantageously be used in the modern system of medicine. It reviews the information about shilajit, its occurrence and uses in traditional and modern system of medicine. It also describes the extraction of novel components, fulvic and humic acids from shilajit and their use as bioavailability enhancers in modern medicine.

It is hoped that the book will stimulate further research in the applications of humic substances in drug development and will be of interest not only to the students and researchers in the field of humic chemistry but also to the researchers and practitioners of traditional and modern medicine as well as to the pharmaceutical research scientists.

The authors are thankful to colleagues, coworkers, students and our families for their support. A special word of appreciation and regards towards Prof. Shibnath Ghosal who has been a pioneer of modern research on shilajit and Dr. D.B.A. Narayana for his inspirational and stimulating thoughts which inspired the present work.

Grateful acknowledgement is made to CSIR, AICTE and UGC for financial support provided for part of the research work carried out by the authors and coworkers and to Jamia Hamdard for providing the facilities.

R. Khanna, S.P. Agarwal & R.K. Khar

ABOUT THE AUTHORS

Dr. Rajesh Khanna is a pharmaceutical research scientist with over 15 years of research experience. He qualified for his B. Pharm., M. Pharm. and Ph.D. in pharmacy from Jamia Hamdard (Hamdard University), New Delhi. He started his career with the Research and Development Centre of J.K. Drugs and Pharmaceutical Ltd. and is currently associated with Dabur Pharma Ltd. specializing in development of dosage forms and drug delivery systems. He has published a number of research papers in national and international journals, has filed a number of patents and has co-authored several books in the field of pharmacy. He takes active interest in professional activities and is a member of several national and international professional bodies.

Prof. Suraj Prakash Agarwal is an academician and pharmaceutical scientist with over 30 years of research experience. He did his B. Pharm. from Birla Instituite of Technology and Science, Pilani and M.Pharm. from Banaras Hindu University, India. He earned Ph.D. from University of Illinois, USA. He has taught pharmacy at University of Illinois, Ahmadu Bello University, University of Nigeria. Since 1988, he has been teaching at Jamia Hamdard, New Delhi and now also at UP Technical University. He has published more than 80 research papers, has filed several patents and has authored a number of books in the field of pharmacy. Dr. Agarwal has been the Head, Department of Pharmaceutics, at University of Nigeria as well as at Jamia Hamdard, New Delhi. He has been a member of selection committees of many universities and an expert member of various statutory bodies of UGC and AICTE.

Prof. Roop Krishen Khar is a renowned academician and a research scientist with more than 30 years of research experience. He obtained

B. Pharm and M. Pharm degrees from Dr. Hari Singh Gour Vishvavidhyalaya (Sagar University) and earned Ph.D. from Faculty of Pharmacy, Medical Academy, Sofia, Bulgaria. Dr. Khar has developed a research school in Controlled Drug Delivery Systems & Pharmaceutical Product Development at Department of Pharmaceutics, Jamia Hamdard, has supervised 55 Ph.D. theses, 70 M. Pharm. theses and published more than 250 research papers in international and national journals. He has several patents to his credit and has also authored a number of text and reference books. He is the Principal investigator of number of research projects funded by various government agencies and pharmaceutical industries in India. Dr. Khar has been the Dean, Faculty of Pharmacy and Head, Department of Pharmaceutics, Jamia Hamdard. He takes very active interest in professional matters and has been the President of Indian Pharmaceutical Association (Delhi Branch). He was awarded the "Teacher of the year award 2002" by Indian Pharmaceutical Teachers Association. He is a member of important selection committees of many Universities, UPSC and an expert member of various statutory bodies of AICTE and Pharmacy Council of India.

CONTENTS

1

HUMIC SUBSTANCES

An Inroduction

Humic substances are heterogeneous mixtures of naturally occurring organic molecules which are ubiquitous in the environment and occur in soils, waters, and sediments of the ecospheres (Schnitzer, 1978; Gaffney *et al.*, 1996). They account for nearly 85% to 90% of the organic matter present in soil; ground water; peat, lignites, brown coals; sewage; and natural waters and their sediments. Humic substances have a relatively high molecular weight and are dark-colored organic materials that are produced during chemical and biological transformation of plant, animal, and human waste (Chefetz *et al.*, 1998). These remarkable biomaterials are crucial components of the carbon cycle and other life processes.

Although the exact chemical structure of humic substances has not been elucidated, it has become apparent that humic substances are not single molecules but rather association of molecules of microbiological, polyphenolic, lignin, and condensed lignin origin. The principal properties of humic acids and their subsequent potential applications depend strongly on their origin (source) as well as on the isolation procedure. The chemistry of humic acids is also deeply influenced by these factors. Recent studies have shown that humic substances have a supramolecular nature in which relatively small heterogeneous molecules are self-assembled by hydrogen bonds, and also by weaker forces such as van der Waals, into large assemblies of only apparently high molecular mass.

Humic substances have been extensively studied for their role in the environment and in agriculture, and excellent reviews are available on the subject (Schnitzer, 1972, 1978; Stevenson, 1994; Gaffney *et al.*, 1996; Tan, 2003). The remarkable properties of humic acids have attracted the attention of many investigators and the results of these studies have pointed to the use of these interesting natural compounds in many practical industrial applications. Humic substances in one form or the other have also been studied and used for their therapeutic properties in animals and humans. The information on this aspect is, however, scattered and not very well documented. The present work tries to collate all this and other remarkable properties of humic substances.

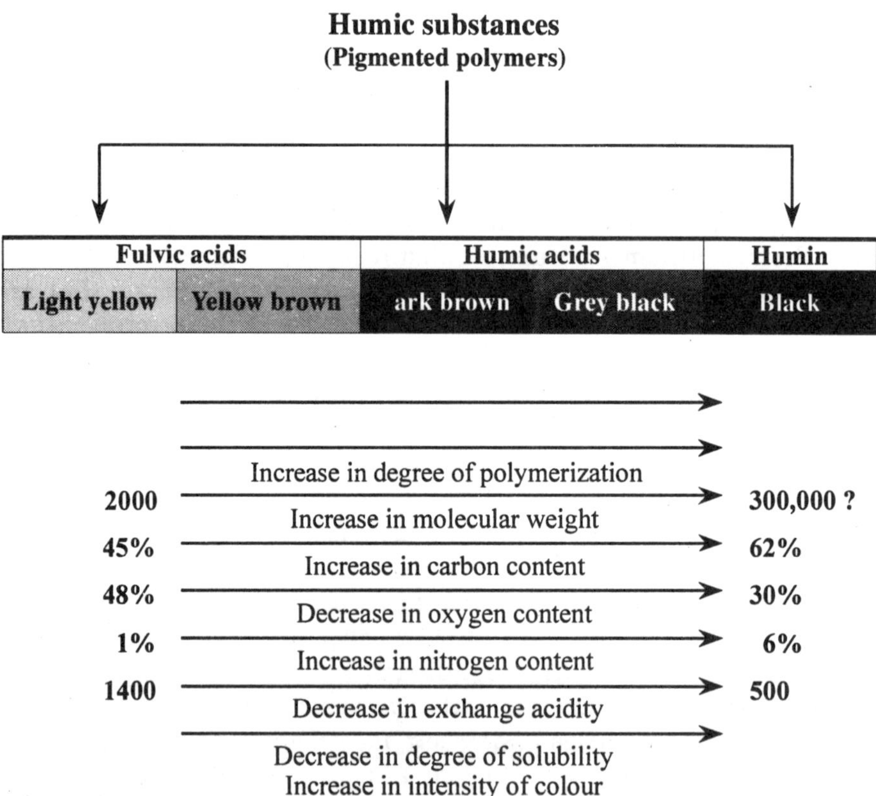

Fig. 1.1: Physicochemical properties of humic substances (Stevenson, 1994)

Humic substances are primarily dark-colored, predominantly acidic, chemically complex, polyelectrolyte, macromolecular materials that range in molecular weight from few hundreds to several thousands. These materials are usually partitioned into the following three main fractions based on their solubility (Fig. 1.1) (Aiken *et al.*, 1985):

Humic acids: The fraction of humic substances that is not soluble in water under acidic conditions (pH < 2), but is soluble at higher pH values. They are dark brown to black in color.

Fulvic acids: The fraction of humic substances that is soluble in water under all pH conditions. These are light yellow to yellow-brown in color.

Humin: The fraction of humic substances that is not soluble in water at any pH value and in alkali. Humin is black in color and is comparatively more aromatic in nature as compared to humic and fulvic acids. It also has comparatively weaker water-retaining, sorbing, and metal-binding properties than fulvic or humic acids. Humin can be considered to be further along in the natural progression from live animals and plants toward "dead" coals and carbon than the fractions, humic acids and fulvic acids (Davies and Ghabbour, 1999).

Humic acids form a major portion of the humus in soils, natural waters, river, lake and sea sediments, brown and brown-black

HUMIC ACIDS: A PROFILE

Synonyms	:	Ulmic acid
Description	:	Dark brown to black amorphous granular powder with a characteristic taste
Solubility	:	Soluble in water above pH 3.0
CAS No.	:	1415-93-6
Elemental composition	:	C = 41–56%, H = 4–6%, N = 13–20%, O = 20–38%
Molecular weight range	:	3000–100,000
Melting point	:	>300 °C

coals, and other natural materials such as shilajit, as a product of chemical and biological transformations of animal and plant residues. Substantial evidence exists that humic acids consist of a skeleton of aryl/aromatic units cross-linked mainly by oxygen and nitrogen groups with the major functional groups being carboxylic acid, phenolic and alcoholic hydroxyls, ketone and quinone groups (Livens, 1991). The large number of diverse chemical functionalities contained in their polymeric nature and relatively high chemical stability favors their practical exploitation (Davies and Ghabbour, 1999) as complexing agents. Being highly aromatic, humic acids become insoluble when the carboxylate groups are protonated at low pH values but become soluble at higher pH values. Humic acids readily complex with metal ions and organics and are anchored by metal binding and by attachment to clays and minerals, which decreases their solubility at a given pH. Humic acid gels retain water very strongly (Davies *et al.*, 1998).

FULVIC ACIDS: A PROFILE

Description	:	Light yellow to yellowish-brown powder with a characteristic taste
Solubility	:	Soluble in water at all pH values
Elemental composition	:	C = 28–39%, H = 4–6%, N = 3–8%, O = 46–62%
Molecular weight range	:	200–3000
Melting point	:	>300 °C

In general, fulvic acids are of lower molecular weight than humic acids, and soil-derived materials are larger than aquatic materials (Stevenson, 1994). The content of C, H, and N is, in general, lower in fulvic acid in comparison with humic acid, while the content of O is comparatively more.

The structures of fulvic acids are somewhat more aliphatic and less aromatic than humic acids and these are comparatively richer in carboxylic acid, phenolic, and ketonic groups than humic acids. This is responsible for their higher solubility in water at all pH values. The total acidities of fulvic acids (900–1400 meq/100 g) are considerably higher than that of humic acids (400–870 meq/100 g).

FORMATION OF HUMIC SUBSTANCES IN NATURE

The formation of humic substances is one of the least understood and the most intriguing aspects of humus chemistry (Stevenson, 1994). Although the formation process of humic substance has been studied in detail and for a long time, their formation is still the subject of long-standing and continued research. Davies and Ghabbour (1999) have provided an excellent review of such theories.

Early studies on the nature and structure of humic substances described humic acid not as a specific compound, but as a mixture of closely related substances having similar structural features and demonstrating the presence of carboxyl groups (COOH) (Aiken *et al.*, 1985). Later studies showed lignin to be the precursor to humic acid. By the mid-20th century it became accepted that humic and fulvic acids were formed by a multistage process that included decomposition of all plant components, including lignin, into simpler monomers; metabolism of the monomers with an accompanying increase in the soil biomass; repeated recycling of the biomass carbon with synthesis of new cells and concurrent polymerization of reactive monomers into high-molecular-weight polymers (Aiken *et al.*, 1985).

Presently, the generally accepted view is that humic substances are produced by the condensation of various components present in the humification process, such as amino acids, lignins, quinones, pectins or carbohydrates, through intermolecular forces (donor–acceptor, ionic, hydrophilic, and hydrophobic), although the mechanisms may vary depending on geographical, climatic, physical, and biological circumstances, respectively. (Pena-Mendez *et al.*, 2005).

Several pathways exist for the formation of humic substances during the decay of plant and animal remains in soil (Fig. 1.2). The classical theory, popularized by Waksman, is that humic substances represent modified lignins (pathway 1) but the majority of present-day investigators favor a mechanism involving quinones (pathways 2 and 3). In practice, all four pathways must be considered as likely mechanisms for the synthesis of humic and fulvic acids in nature, including sugar-amine condensation (pathway 4) (Stevenson, 1994).

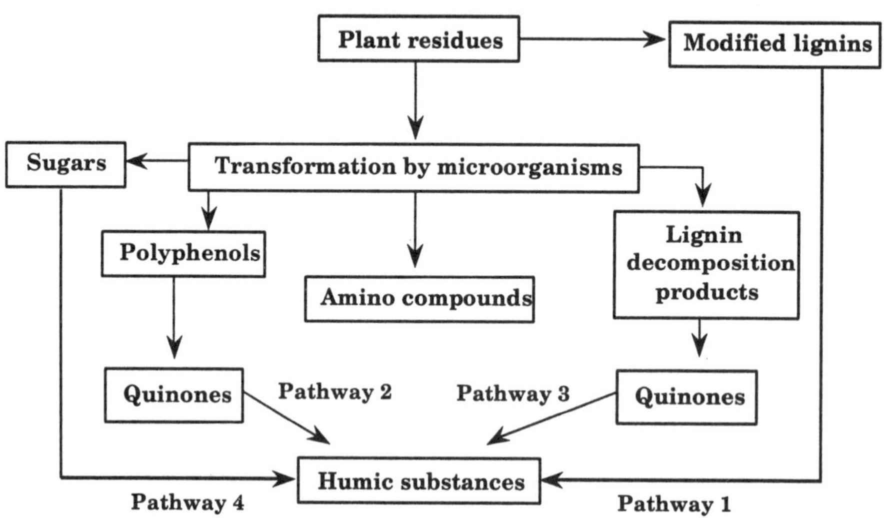

Fig. 1.2: Mechanisms of formation of humic substances (Stevenson, 1994)

MOLECULAR AND CHEMICAL STRUCTURES OF HUMIC SUBSTANCES

The structures of humic substances (humic and fulvic acids) have not been completely elucidated despite many decades of research, although a number of tentative structures have been proposed by various workers. The main task that confronts researchers in this field today is to develop a valid concept of the chemical structure of humic materials. Humic substances are believed to consist of molecules ranging in molecular weight from a few hundreds to several hundred thousands. Individual fractions such as humic and fulvic acids merely represent a particular part of this molecular-weight range (Livens, 1991). Nevertheless, since they are derived from chemically similar starting materials, all the molecules of humic substances share some structural and chemical characteristics.

A number of techniques such as NMR (Frund and Ludemann, 1989; Simpson *et al.*, 2002), mass (Novotny and Rice, 1995), X-ray (Rice *et al.*, 1999), spectroscopy (Spiteller and Schnitzer, 1983; Shin *et al.*, 1999), and a number of other allied techniques (Grasset and Ambles, 1998; Avena *et al.*, 1999) have been employed to investigate the detailed structure of humic substances. Figure 1.3 shows a hypothetical model structure of humic acid proposed by Stevenson (1994).

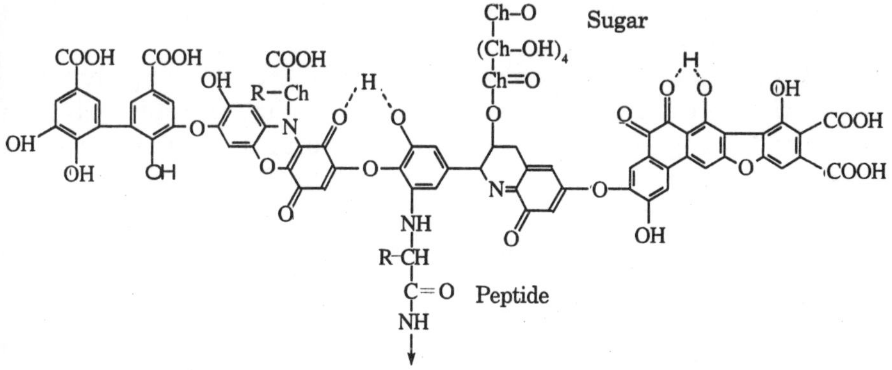

Fig. 1.3: Model structure of humic acid proposed by Stevenson (1994)

In the broadest terms, the structures can be described as assemblies of covalently linked aromatic and aliphatic residues carrying carboxyl, phenolic, and alkoxy groups cross linked mainly by oxygen and nitrogen groups, although sulfate esters, alanine moieties, semiquinone, phosphate ester and hydroquinone groups have been proposed to exist in some humic isolates. With time, it has become more apparent that humic substances are not single molecules but are rather association of molecules of microbiological, polyphenolic, lignin, and condensed lignin origin and no single structural formula will suffice to describe them. X-ray analysis and viscosity measurements of humic substances have shown them to have an "open" flexible structure perforated by voids of varying dimensions that can trap or fix organic or inorganic compounds like carbohydrates and proteins, besides others, that fit into the voids provided that the charges are complimentary (Schnitzer, 1978).

Studies carried out using multidimensional NMR have presented the structure of humic substances as macromolecular aggregate composed from mixture of relatively low-molecular-weight (<2 kDa) organic compounds (aliphatic acids, ethers, esters, and alcohols, aromatic lignin derived fragments; polysaccharides and polypeptides) holding together through a complex combination of hydrophobic, charge-transfer, and hydrogen bond interactions, and metal bridging (Simpson *et al.*, 2002).

According to another recent view of the molecular structure of humic substances, these can be described as collections of diverse and relative low molecular weight components forming dynamic associations stabilized by hydrophobic interaction and hydrogen bonds and being capable of organizing into micellar structures in suitable aqueous environment (Sutton and Sposito, 2005).

COLLOIDAL CHARACTERISTICS OF HUMIC SUBSTANCES

The colloidal state represents a phase intermediate between true solution, where species are of ionic or molecular dimension, and suspended particulates, where species are sufficiently large to settle under the force of gravity. The colloidal range is considered to extend from 0.001 to 1 mμ. Chemical and physical reactions are generally enhanced in colloidal systems due to the large surface area of colloidal particles. The ranges of molecular size for the majority of humic and fulvic acids place them in the colloidal range when in aqueous solution (Gaffney *et al.*, 1996).

Humic colloidal material is thought to consist of coiled, long-chained or three-dimensional cross-linked macromolecules with electrical charges variously distributed on the particle. The presence of charged sites arising from ionized groups results in mutual repulsion and causes maximum expansion of the macromolecule (Stevenson, 1982). The factors most important in controlling the molecular conformation of humic materials are concentration of the humic materials, pH, and ionic strength of the system. At high sample concentration (>3.5 g/L), low pH (<3.5), and high electrolyte concentration (>0.05 M), the humic materials are rigid uncharged colloidal particles. At low concentrations, high pH, and low electrolyte concentrations, humic and fulvic acids exist as flexible linear polyelectrolytes.

ROLE OF HUMIC SUBSTANCES IN ENVIRONMENT

Humic substances are surface active, by virtue of the hydrophobic and hydrophilic moieties coexisting in a single molecule, and have a tendency to form micelles in solutions at and above their critical micelle concentration. This gives them the ability to play important role in the solubilization and transportation of hydrophobic chemical entities in nature. The presence of humic material can also promote the solubilization of non-polar hydrophobic compounds. Such ability can affect not only the mobility of the bound molecule but also the rate of chemical degradation, photolysis, volatilization, and biological uptake (Gaffney *et al.*, 1996).

In the environment, humic acids can bind metal ions from dissolved rocks and minerals, can interact with other soil components such as clay particles, and can bind pollutants and biocides used for agricultural purposes. Humic acids play a major role in the geocycling of metal ions and in the transport of pollutants and biocides in the environment. It has also been frequently suggested that humic substances play a major role in controlling the behavior and mobility of metals in the environment by forming complexes with them (Livens, 1991).

Humic substances, especially humic and fulvic acids, present in soil are known to increase nutrient uptake, drought tolerance, and seed germination in plants. They increase the availability of nutrients that are already in the soil and naturally aerate the soil from inside. Researchers have also recognized their ability to complex metals and radionuclides and to interact with free radicals.

2

TRADITIONAL USES OF HUMIC SUBSTANCES IN FOLKLORE MEDICINE

There are a number of natural humic substances which have been identified and used traditionally as part of oriental medicine worldwide. The usage has evolved from centuries of knowledge sharing, careful observation, and practical experience gained from using these amazing bio-products of nature. One of the earliest documented use of humic substances dates back to about 1000 BC when the ancient Indian Ayurvedic texts of *Charaka Samhita* and *Sushruta Samhita* described shilajit, an exudate from the Himalayan rocks as a rasayana (rejuvenator) as well as a panacea (cure) for all diseases and endowed with a capability of increasing the human longevity (Acharya, 1962; Sharma and Das 1988; Ghosal *et al.*, 1995e; Agarwal *et al.*, 2007). A similar rock exudate from the Caucasus, Altai, and Pamir mountain ranges has been used for over 3000 years as part of folklore medicine in the former Soviet Union, China, Tibet, and Greece (Frolova and Kiseleva 1996; Schepetkin *et al.*, 2002). It has been known by the name of Mumie or Mumio or Mumijo in these countries and has been quite popular for the treatment of bone fractures, dislocations, diseases of the skin and the peripheral nervous system and also as soothing and anti-inflammatory agent. Greek physicians used it as an antidote to poisons and in the treatment of arthritis and inflammation. Aristotle studied the medicinal effects and applications of Mumio and its medicinal use was promoted and spread by his pupil, Alexander the Great, throughout his empire.

Galen included Mumio as an ingredient in his famous panacea, Theriac. The Persian physician Al Biruni recommended Mumio for traumatic injuries. The two medieval medical giants Avicenna and Paracelsus both knew of and used Mumio. Humic substances have also been used for centuries in Germany, Poland, and other European countries where the products are usually derived from peat deposits. Mud therapy has been used since long in Babylonia and the Roman Empire, where the inhabitants already recognized the healing effects of mud (Klocking and Helbig, 2005).

SHILAJIT OR MUMIE

Shilajit, also known as salajit, shilajatu, mumie, or mummiyo is a pale-brown to blackish–brown exudation, of variable consistency, coming out from layer of rocks in many mountain ranges of the world, especially the Himalayan and Hindukush ranges of the Indian subcontinent (Chopra *et al.*, 1958; Ghosal, 1992; Agarwal *et al.*, 2007). It is also found in Australia, Bhutan, China, Egypt, Mongolia, Nepal, Norway, Pakistan, Russia, and other countries, where it is collected in small quantities from steep rock faces at altitudes between 1000 and 5000 m. In its raw form, shilajit is a semi-hard, brownish-black to dark, greasy, black resin having a distinct smell and taste. Shilajit samples from different region of the world, however, vary in their physiological properties.

Fig. 2.1: Rock shilajit in its raw form

Although the name shilajit means "Winner of Rock", shilajit itself is not a rock comprising inorganic material but is a complex mixture of organic humic substances and plant and microbial metabolites occurring in the rock rhizopheres of its natural habitat. Figure 2.1 shows the photograph of shilajit in its raw form.

SHILAJIT: IN ANCIENT TEXTS

Shilajit finds mention in a number of ancient texts of the Ayurvedic and Unani systems of medicine written about 1000 years BC to 14th Century AD, most notable being the Ayurvedic medical treatises of *Charaka Samhita* and *Sushruta Samhita* (Ghosal *et al.*, 1995e). These ancient Sanskrit texts have described shilajit as a rejuvenator and an adaptogen endowed with miraculous healing properties.

Shilajit: Synonyms

Shilajit has various synonyms (Chopra *et al.*, 1958; Nadkarni, 1976; Frolova and Kiseleva, 1996) (Table 2.1). In Sanskrit, it is called Shilajatu, Shilaras, Silajit, Adrija, or Girija (meaning derived from rocks). In English, it has been known as asphalt, bitumen, mineral pitch, or Jews pitch. In Hindi, Gujarati, and Marathi, it is called, Shilajit or Silajita. In Persian, it is called, Momiai Faqurul Yahud; in Russian, it is known as Mummiyo or Myemu; and in German, it is known as Mumie or Mumijo which in Greek means "saving body" or "protecting organism". The Tibetans call it Brag-shun meaning "mountain oil", while in Arabic, it is called Hajar-ul-musa or Arakul dzibol meaning "mountain sweat".

The Sanskrit meaning of shilajit is "conqueror of mountain and destroyer of weakness". Other terms like dathuras, dathusara, and shiladhatu have also been used in ancient medical texts like *Sushruta samhita*, *Charak samhita,* and *Rasarangini* to describe shilajit. The word dhatu has been used simply to emphasize its capability as a rasayana, which increases the activity of the saptadhatus of the body (Tewari *et al.*, 1973).

Table 2.1: Vernacular names of shilajit

Language	Vernacular name
Arabic	Hajar-ul-musa, Arakul dzibol
Burmese	Kaotui
English	Asphalt, bitumen, mineral pitch, Jew's pitch
German	Mumie
Greek	Mumijo
Indian	
Bengali	Silajatu
Gujarati and Marathi	Silajita
Hindi	Shilajit, Silajit, Ral-yahudi
Sanskrit	Shilajit, Shilajeet, Shilajatu, Shilaras, Silajit, Adrija, or Girija
Tamil	Perangyum, Uerangyum
Kyrgyz	Arkhar-tash, Momiya; Momlai
Latin	Asphaltum
Mongolian	Brag-shun
Persian	Momiai Faqurul Yahud
Russian	Mummio, Mummiyo, Momiyo, Myemu, Mumie
Siberian	Barakhshin
Tajik	Zogh, Kiem
Tibetian	Brag-shun
Turkmen	Mumnogai
Ujbek	Asil

VARIETIES OF SHILAJIT

There are four different varieties of shilajit which have been described in *Charaka Samhita*, *viz*., savrana, rajat, tamra, and lauha shilajit (Chopra *et al*., 1926). Savrana shilajit is gold shilajit and is red in color. Tamra shilajit is copper shilajit and is blue in color. Rajat shilajit is silver shilajit and is white in color, while lauha shilajit is an iron containing shilajit and is brownish-black in color. Tamra and Savrana shilajit are not found commonly but the last variety, *i.e*., lauha shilajit is commonly found in Himalayan ranges and is supposed to be the most effective according to the therapeutic point of view. If administered according to proper procedure, it produces rejuvenating and aphrodisiac effects and cures diseases (Sharma and Das, 1988). There is also a mention of karpura shilajit or white shilajit in the literature.

Sushruta Samhita mentions six types of Shilajit based on their origins. In addition to the four types listed above, there is a mention about tin and lead shilajit. Each type has the same taste (rasa) and potency (virya) as the metal to whose essence it owes its origin. According to *Sushruta Samhita,* as tin, lead, iron, copper, silver, and gold are progressively more efficacious, the different types of shilajit that derive from these metals are also progressively more efficacious in their application (Bhishagratna, 1998).

Based on its features and origin, Russian Mumie has been categorized into three types: petroleum mumie, plant mumie (mumie-asil), and mumie-kiem (Schepetkin *et al*., 2002). Petroleum mumie is believed to be formed as a result of transformation of deep petroleum products of mountains. Mumie-kiem is assumed to be resulting from the long-term humification of guano (feces) of alpine rodents, while Mumie-asil is believed to be formed due to the long-term humification of Euphorbia and Trifolium (clover) plants and lichen. Of the three varieties, Mumie-asil is believed to have the highest therapeutic quality (Schepetkin *et al*., 2002).

SHILAJIT: OCCURRENCE AND DISTRIBUTION

Shilajit has been found to be widely distributed in several mountain ranges of the world. In India, it is widely distributed in the Himalayan range from Arunachal Pradesh in the east to

Kashmir in the west. Other mountain ranges of the Indian subcontinent where it is found distributed include the Hindukush ranges, the Aravallis, the Vindhyas, the Eastern Ghat, and the Western Ghat ranges (Ghosal *et al.*, 1976, 1988; Ghosal, 1989, 1990). Shilajit is also found widely distributed in the mountain ranges of countries like Afganisthan (Badakh-shan), Australia (Northern Pollock), Bhutan, China (Yunnan), Nepal (Dolpa), Pakistan (Peshwar), Russia (Altai, Caucasus, Tien-shan, Ural), Tajakistan, Uzbekistan and USA (the Grand Canyon) (Ghosal *et al.*, 1988; Ghosal, 1990), although it may be referred to by any of its vernacular name at different locations.

During the summer months, when the weather is hot in the mountains, shilajit melts and trickles down the crevices and spreads on the rock surface from where it is collected in small quantities. It is also found embedded in the interior of the rock sediments from where it gets exposed, in association with clay minerals, after landslides, geological excavation, or mountain road cutting (Chopra *et al.*, 1958; Ghosal *et al.*, 1976, 1989; Kong *et al.*, 1987). Being a natural substance, the composition of shilajit varies from place to place and is influenced by factors such as the plant species involved, the geological nature of the rock, local temperature profiles, humidity, and altitude.

ORIGIN OF SHILAJIT

There are a number of hypothesis about the origin of shilajit (Tiwari and Agarwal, 2002). Shilajit is believed to have been derived from vegetation fossils that have been compressed under layers of rocks for hundreds of years and have undergone a high amount of metamorphosis due to the high-temperature and pressure condition prevalent there. During warm summer months, shilajit becomes less viscous and flows out between the layers of rocks. Ancient Indian texts of *Sushruta Samhita* and *Rasarangini* also suggest that shilajit has a vegetative origin. It has been mentioned in *Sushruta Samhita* that in the month of May–June, the sap or latex juice of plants comes out as a gummy exudation from the rocks of mountain due to strong heat of sun. Ancient Indian texts of *Rasarangini* and *Dwarishtarang* also claim that shilajit is an exudation of latex gum resin, etc., of plant, which comes from the rocks of mountain under the presence of harsh scorching heat (Tiwari *et al.*, 1973).

SHILAJIT: USES IN TRADITIONAL SYSTEMS OF MEDICINE

Shilajit has an important and unique place in the traditional texts of Ayurveda, Siddha, and Unani medicine. In regional folk medicine, shilajit is a reputed rasayana (a rejuvenator and immunomodulator), claimed to arrest the process of ageing and prolong life (Phillips, 1997). It is prescribed for the treatment of genitourinary disorder, jaundice, gallstone, digestive disorders, enlarged spleen, epilepsy, nervous disorder, chronic bronchitis, and anaemia (Chopra *et al.*, 1958). Shilajit has also been known to be useful for treating kidney stones, edema, piles, internal antiseptic, adiposity, to reduce fat, and anorexia (Nadkarni, 1976). Shilajit is given along with milk to treat diabetes. Shilajit is prescribed along with guggul to treat fracture. It is believed that it goes to the joints and forms a callus quickly. The same combination is also used to treat osteoarthritis and spondylosis.

Shilajit has also been ascribed potent aphrodisiac property. The Indian Ayurvedic System of Medicine recognizes health as a unification of body, mind, soul, and sense organs. It postulates that all the four factors should be in optimum condition to obtain a better health. Thus, it is imperative for an individual to maintain his body, mind, soul, and senses in order to be in a better performing state. Similarly, Ayurveda looks at lack of libido, through the problems associated with mind and body. These include anxiety neurosis, depression, mental and physical stress, mental and physical fatigue, lack of proper digestion, lack of smooth functioning of heart (blood circulation), liver, etc. As a matter of fact, Ayurveda addresses all these root causes to correct the lack of libido. Hence it recommends "Rasayana" which do function on all these systems and correct the imbalance thus giving an overall well-being. It is postulated that once the physical body as well as the mind becomes healthy, the libido is bound to be normal (Dabur Shilajit, 2007). Shilajit is perhaps the best Rasayana, Ayurveda has prescribed. The underlying power of Shilajit comes from its potency of reducing the stress of internal organs, effecting balanced energy metabolism, and anti-ageing. It rejuvenates and heals every body cell and helps the cell grow at any age. Obviously, a younger body can withstand all types of ailments and can better perform.

According to Ayurveda, shilajit arrests the process of aging and produces rejuvenation which are two important aspects of an Ayurvedic rasayana. Shilajit has been used widely in combination with honey for a number of applications. When this usage is seen in context of modern scientific understanding, it becomes apparent that the combination had both practical and therapeutic reasons. On one hand, honey could pretty well mask the pungent taste of Shilajit and on the other hand, the user gets benefited from the advantages of honey therapy combined with the beneficial effects of shilajit.

The Ayurvedic energenies vary depending on the base rock that the shilajit comes from, but it is generally thought to be *tridoshic* and only aggravating to *Pitta* (fire) when used in excess. *Rasa*, *Guna*, *Virya*, *Vipaka*, *Dravya,* and *Prabhawa* are the criterias on the basis of which medicinal value of any substance is considered in Ayurveda. The *Charaka Samhita* gives the ayurvedic energies for each type of shilajit as summarized in Table 2.2.

Table 2.2: Ayurvedic energies for different types of shilajit

Shilajit type	Color	Rasa	Virya	Vipäka
Gold	Similar to flower of japä (hibiscus)	Bitter	Cooling	Pungent
Silver	White	Pungent	Cooling	Sweet
Copper	Peacock throat	Bitter	Hot	Pungent
Iron	Similar to gum of guggulu	Bitter, salt	Cooling	Pungent

Mumie has traditionally been used in Asian herbal medicine, systemically and topically, for the treatment of bone fractures, dislocations, diseases of the skin, diseases of the peripheral nervous system (neuralgia and radiculitis), and also as a soothing and anti-inflammatory agent. Momio was extensively used by *hakims* (practitioners of Unani system of medicine) in the treatment of inflammatory and genitourinary disorders, and as an aphrodisiac. Myemu is used in Russia and the other Republics, in the treatment of diabetes, arthritis and to slow down the process of aging, while mumie currently finds a place in phytotherapy in Germany for treating a number of ailments. Greek physicians used it as an antidote to poisons and in the treatment of arthritis and inflammation.

PURIFICATION AND FORMULATION OF SHILAJIT

Crude shilajit in its natural form is often contaminated with varied amount of impurities such as mycotoxins, heavy-metal ions, polymeric quinones, reactive free radical, microbial toxins/cellular debris of animals, and microorganisms, etc. (Ghosal *et al.*, 1995c). In a study on crude Shilajit from Kumaon region of India, the rhizosphere of shilajit samples was found to be heavily infested, at the periphery with fungal organisms (Table 2.3) (Ghosal *et al.*, 1991; Kumar, 1993). Based on population percentage, one fungal form *Aspergillus niger* was found to be present in almost 100% of the samples, while three forms – *Cladosporium oxysporum*, *Trichothecium roseum,* and *Ulocladium chartarum* were present in almost 75% of the samples. It is interesting to note, however, that no bacterial association could be seen as against soil humic samples which are most often associated with bacterial infestations (Ghosal *et al.*, 1991).

Mycotoxins are produced by mold or fungi and can cause illness or death in man. Free radicals can be harmful to cells and are believed to be a causative factor in aging. Polymeric quinones are an oxidation product of quinic acid which is found in some plants. Hence, it is necessary to purify shilajit before it is consumed. The findings are consistent with the ancient texts which recommend purification of shilajit before consumption.

Shilajit: *Shodhana*

Detailed procedures have been prescribed in the original Ayurvedic texts for graded purification (*Shodhana*) and subsequent formulation (*Bhavana*) of shilajit. According to these recommendations, it is necessary to remove first the water insoluble inorganic and polymeric materials from the external surface (*Bahirmal*: external impurities) of shilajit. Washing with water is sufficient for it. To remove the internal impurities (*Antamal*: internal impurities) from the void/pores of shilajit would require neutral salts (sodium pyrophosphates, ammonium chloride)

Table 2.3: Fungal organism and their respective population from shilajit rhizosphere (Ghosal *et al.*, 1991; Kumar, 1993)

Fungal organisms	Population occurrence (%)
Aspergillus niger	100
Cladosporium oxysporum	75
Trichothecium roseum	75
Ulocladium chartarurn	75
Aspergillus sydowii	50
Aspergillus ochraceus	50
Alternaria alternata	50
Absidia corymbifera	50
Botryodiplodia theobromae	50
Fusarium equisiti	50
Fusarium compactum	50
Fusarium moniliforme, var, *subglutinans*	50
Hormoconis resinae	50
Mrimbla ingelheimense	50
Pestalotiopsis sp.	50
Sporomiella isomera	50
Aspergillus ustus	25
Penicillium aurantiogriseum	25
Rhizoctonia sp.	25

and buffers (citrate, lemon juice) of mineral and organic acids. The remaining part of the impurities such as loose metal ions, polymeric quinones, and other free radicals can then be removed by treatment with small tannoids such as from pistacia species or with *Triphala* (Ghosal *et al.*, 1995e).

In another method, raw shilajit rock powder is first exhaustively extracted with water and then filtered to remove the water insoluble impurities. The filtrate is then concentrated either by direct heating to produce a substance known as *agni-tapi* (heated by fire) – shilajit; or by keeping the filtrate under direct sunlight to give *surya-tapi* (heated by sunlight) – shilajit. The creamy layer that is formed during heating process is skimmed three to four times to get pure shilajit (Ghosal *et al.*, 1995e).

Shilajit: *Bhavana*

Ayurvedic texts also describe the process of *Bhavana* or impregnation of shilajit. According to this method, the micropores of shilajit fulvic acid are made vacant by the previously described process of purification (*Shodhana*). These empty micropores are then filled up with other bioactive molecules (drugs) according to the requirement of the specific disease. The varieties of shilajit which are sufficiently rich in bioactive substances (*e.g.*, oxygenated dibenzo-α-pyrones) are called as *Satwajacta* shilajit and need not be used for the process of *Bhavana*/formulation. It is the *Nisatwa* shilajit whose fulvic acid micropores have enough vacant spaces, which should be used for the process of *Bhavana* (Ghosal *et al.*, 1995e). Different traditional formulations of shilajit are prepared by further addition of aqueous extracts of medicinal plants such as *Emblica officinalis, Terminalia chebula, Terminalia belerica, Tribulus terrestris, Aegel marmelos, Gmelina arborea, Solanum indicum,* and *Solanum xanthocarpum* individually or in combination to the purified shilajit to produce potent shilajit formulations for treating different ailments and diseases (Ghosal, 2006). This process of using additional extracts to purified shilajit is known as the *Marna* (potentization by impregnation of medicinal agents).

MARKETED PREPARATIONS OF SHILAJIT

A number of formulations of shilajit are marketed by pharmaceutical companies in India (Dabur, Arogya Herbals, Baidhyanath, Indian Herbs, Gurukul Kangri), the United States (Fabri-Chem, Triple Crown, Natreon), Pakistan, Ukraine, Kazakhstan, Russia and a number of other countries. These formulations usually contain purified or shodhit shilajit either alone or in combination with other herbal and mineral components. The formulations are either in the form of thick paste or as powder filled into hard gelatin capsules. Some of these preparations have resulted from the development of Ayurvedic conceptions, the others were devised by contemporary research and patented. Some of the formulations of shilajit marketed world-wide include Dabur Shilajit capsules (Dabur), Shilajit Gold capsules (Dabur), Shudh Shilajit (Gurukul Kngri), Shilagen (Healthy Outlook Nutrition), Primavie (Natreon), Abana, Cystone, EveCare, Gericare, Stresscare, Geriforte, Lukol, Pilex, Rumalaya, Tentex forte and Nefrotec (Himalaya), Adrenotone (Rockwell Nutrition), Siotone (Albert David), La-Tone Gold (La Medicca), Andro-Surge (Mineral Connection), Libidoplex (Solanova.) and Renone cream (Yogi Herbals).

Dabur Shilajit

Dabur shilajit capsules contain purified extract from 500 mg of raw shilajit and are indicated to be taken twice daily, preferably with milk, as a health tonic. Shilajit has been known to support immune system, optimize physical performance, enhance rehabilitation of muscle, bones and nerves and help in acclimatizing to newer environmental and climatic conditions (Dabur Shilajit, 2007). *Shilajit Gold* is a unique combination of Shilajit, Gold, Kesar and other important herbs like Ashwagandha, Kaunch Beej and Safed Mushali. Shilajit Gold is claimed to be a powerful sex stimulant, which improves sexual health and acts as a rejuvenator to increase stamina, vigour and vitality (Shilajit Gold, 2007).

Shilagen

Shilagen is a unique formulation containing shilajit along with proven healing substances such as ashwagandha, ginkgo biloba, bacopin, and trace minerals to aid in the absorption and synergy of the primary ingredient, shilajit. The producers of Shilagen use a patented oxygen/nitrogen displacement extraction process that ensures the proper pH and increases the effectiveness of active ingredients of shilajit by approximately 800%. They also use a standardized extract, so equal high levels of active ingredients are in each bottle. Shilagen has been recommended for the treatment of the same disorders for which shilajit extract is applicable (Schepetkin *et al.*, 2002).

PrimaVie

Natreon Inc.'s patented ingredient, PrimaVie is a purified and standardized shilajit extract for nutraceutical use to help revitalize the body and mind. PrimaVie is standardized to have not less than 60% fulvic acid and equivalents with high levels of dibenzo-α-pyrones and dibenzo-α-pyrone chromoproteins. These carefully elucidated bioactive components are claimed to assist in the maintenance and regeneration of normal physiological functions by acting as an energy currency, or biocatalyst, in the body. As an adaptogen, PrimaVie is claimed to help the body increase resistance to fatigue while providing energy, vitality, and well-being by augmenting coenzyme Q-10 activity (Primovie, 2008).

Adrenotone

Adrenotone is a polyherbal formulation distributed by Rockwell Nutrition Company and Gaines Nutrition and contains shilajit in combination with Chinese herbs used for centuries (Schepetkin *et al.*, 2002). Adrenotone has been recommended as a remedy to naturally support the adrenal function, high energy levels, and overall wellness during times of stress or immune weakness (Ziauddin *et al.*, 1996).

Siotone

Siotone is a herbal formulation comprising of *Withania somnifera*, *Ocimum sanctum*, *Asparagus racemosus*, *Tribulus terristris* and shilajit, all of which are classified in Ayurveda as rasayanas which are reputed to promote physical and mental health, improve defence mechanisms of the body and enhance longevity (Bhattacharya *et al.*, 2000).

La-tone Gold

La-tone Gold is a composite Herbal formulation useful for sexual impotence in males. Composed of shilajit and several traditional herbs and minerals, La-tone Gold provides a rapid arousal by exerting a generally stimulating and enhancing influence on functions related to the reproductive system that promote enhanced genital blood circulation and engorgement, emporia and a satisfying tumescence besides improving stamina in males (La-tone Gold, 2008).

Andro-Surge

Andro-Surge from Mineral Connection is an herbal formulation designed for optimal regulation of anabolic hormones and testosterone levels.

This formula is recommended for male athletes, being especially effective for those over 38 years of age, or for adults with low levels of dehydroepiandrosterone (Schepetkin *et al.*, 2002).

Libidoplex

Libidoplex is a synergistic blend of specific herbal extracts and vitamins. It is claimed to be a powerful virility-enhancing formula designed to height-en energy and stamina. It can also be used as an overall body tonic (Solenova libidoplex, 2008).

Renone

Renone cream is an external application ointment formulated for greater relief from muscular pains, joint pains, swelling, and inflammation. It is an exclusive combination of many reputed and time-tested pain-relieving herbs and ingredients including shilajit whose antioxidant and anti-inflammatory properties are claimed to decrease and relieve joint inflammation and pain (Renone, 2008).

Other Polyherbal Formulations

Abana, Cystone, Eve Care, Geriforte, Lukol, Pilex, Rumalava, Tentex forte and Nefrotec are some of the other polyherbal formulations containing shilajit in combination with herbal extracts of proven therapeutic efficacy (Shilajeet, 2008). Evecare is a herb-mineral uterine tonic formulated by Himalaya Drug Company. The herbs used in this tonic are effective in various menstrual disorders. These components acting alone and in combination are responsible for the efficacy of the drug in dysmenorrhea, menorrhagia, and other uterine disorders (Mitra *et al.*, 1998). Shilajit is one of the components of Stress Care and Geri Care. Stress Care is useful in stress-related conditions such as premature aging, fatigue, insomnia, or emotional imbalance. Geri Care is the ultimate general fitness product that promotes health and helps everyone to age gracefully without any adverse effects. Geriforte is a completely natural product that regulates and balances all the body organs and systems for comprehensive health care maintenance. Geriforte is used as a restorative tonic to solve the problems of old age in India.

NEED FOR STANDARDIZED FORMULATIONS OF SHILAJIT

As previously mentioned, crude shilajit in its raw form is not suitable for direct human administration due to the presence of mycotoxins, free radicals and polymeric quinones. In addition, owing to its natural origin, crude samples from different locale vary both qualitatively and quantitatively, with respect to their active principles as well as carrier molecules contained therein. This can in turn result in marked variation in the biological effects of shilajit obtained from different sources (Ghosal *et al.*, 1991).

A possible solution could be to purify and standardize formulated shilajit products on the basis of their active principles and carrier molecules to obtain the best possible therapeutic benefits from this panacea of oriental medicine. This would also help in titrating the optimum dose which would provide the desirable therapeutic effect while being devoid of adverse side effect.

OTHER THERAPEUTICALLY USED HUMIC SUBSTANCES

Humic substances of medicinal importance are found abundantly in peat, sapropel, and other natural humified sources. Peat is organic soil formed as a result of incomplete disintegration and humification of died marsh plants in conditions of high humidity. The organic matter of peat in 90% consists of humin, humic and fulvic acids (up to 40%), lignin, polysaccharides, lipids, pectines, hemicellulose, and cellulose (Schepetkin *et al.*, 2002). Sapropels are silted deposits of water reservoirs (lakes, peat marshes, sea estuaries) and contain a large quantity (450%) of organic matter (lignin-humus complex, carbohydrates, bitumen, etc.) in colloidal state. The organic components of "mature" sapropel are produced by the slow decomposition and humification of plants and phytozooplankton in anaerobic conditions. This process is accompanied by the condensation of phenolcarboxylic acids with formation of new high-molecular organic compounds such as humic and fulvic acids. Peat and sapropel are used in pelotherapy as external remedies.

The humification process leads to change in pharmacological properties of peat extracts, in parti-cular, to enhancement in antiulcerogenic and antiradical activity of peat extracts. For some peat humates, the antitoxic properties are also characteristic. Pharmacological properties of peat extracts are studied mainly on such patented drugs as Torfot (Russia) and Tolpa Peat Prepa-ration (TPP) (Poland). The chemical composition of these preparations is standardized which makes it possible to conduct systematic studies on the influence of peat extracts on biological systems.

Peat and sapropel are used in pelotherapy as external remedies (Schepetkin *et al.*, 2002). Two examples of patented peat extracts are

Torfot (Russia) and Tolpa Peat Preparation (Poland). TTP is used in Eastern European countries for a variety of therapeutic purposes, primarily as an immuno-modulator.

In Chinese medicines, fossilized or partly fossilized resins, such as amber and *hutonglei,* are believed to bear close resemblance to shilajit (Kong *et al.,* 1987). The resins of amber, derived from conifers and that of *hutonglei,* derived from members of Salicaceae, after staying in soil for a long period of time are transformed into humus – like substances.

Torfot

Torfot is a product of distillation of specific peat layers. As a medicinal drug, it is a sterile liquid having a characteristic smell of peat. Torfot is administered in the form of hypodermic or subconjunctival injections in ophthalmology for treatment of patients with keratitis, chorioretinitis, and vascular and degenerative processes in the retina (Schepetkin *et al.*, 2002). Torfot possesses antibacterial and antiinflammatory action, and improves blood circulation and tissue regeneration. These properties of the drug are the reason for its application in stomatology. It was shown that Torfot is also applicable for complex treatment for other chronic inflammatory diseases and pulmonary tuberculosis.

Tolpa Peat Preparation (TPP)

Tolpa Peat Preparation is a peat derived pharmaceutical product manufactured by Torf Corporation, Poland. It has immunocorrecting properties and is available in the forms of Tolpa tablets, Chamosaldont Gel (a dental product), and Hypocalen Gel (a dermatological product) (Tolpa Peat Preparation, 2008) . These medications are manufactured on a base of peat extract which is obtained from selected peat deposits in ecologically clean and unpolluted areas in Poland. In addition to other ingredients, TPP contains a wide variety of amino acids and micro- and macro-elements. Scientific studies of the peat extract have been conducted over many years and they confirm the efficiacy of peat as immuno-modulatory and therapeutic agent. Extensive toxicity studies carried out on the product have shown no embryotoxic or teratogenic effects in hamsters or rats after administration of daily doses ranging from 5 to 50 mg/kg (Schepetkin *et al.*, 2002). TPP was found to be neither mutagenic nor genotoxic in selected short-term tests, and was unable to induce or enhance an allergic sensitization in mice and guinea. TTP is effective in the treatment of inflammatory conditions of the cervix, especially cervical erosions. It has also been found useful in cases of recurrent respiratory tract infections and for the treatment of duodenal and peptic ulcers.

Sapropel

Sapropel (from the Greek words *sapros* and *pelos*, meaning putrefaction and mud, respectively) is a term used in marine geology to describe dark-coloured sediments that are rich in organic matter. These are silted deposits in water reservoirs, *e.g.* lakes, peat marshes, sea estuaries, which contain a large quantity of organic matter in a colloidal state. The organic components are produced by slow decomposition and humification of plants and phytozooplankton in anaerobic conditions. Sapropels of different kinds are varied in their ability to correct hepatic function in rats with toxic hepatitis and have positive effect during experimental therapy of pancreatitis (Schepetkin *et al.*, 2002). The restoring influence of sapropels appears to occur due to both their adaptogenic and antioxidant effects. At present, the medicinal remedies Peloidodistillate, Humisol, Peloidin, and FiBS are produced from the different kinds of sapropels.

Peloidodistillate

Peloidodistillate is produced by distillation of sapropel from Tambukan lake (Caucasian region, Russia). The marketed preparation, Vitapeloid is a 1% solution of pyridoxine hydrochloride in Peloidodistil-late. Therapeutic effect of the drug is caused by the presence of phenolcarboxylic acids, amines, vitamins, and microelements in its composition. The preparations stimulate metabolic processes in organism, accelerate regeneration ability, increase the organism resistance to unfavorable factors, and activate immunity. The drug does not possess allergic, teratogenic, and carcinogenic properties. It is applied in ophthalmology for treating patients for degenerate processes of cornea and retina, and initial forms of optical nerve atrophy (Schepetkin *et al.*, 2002). The drug is also recommended against radiculites and neuralgias. In gynecology, it is used against chronic inflammatory processes.

Humisol

Humisol is a 0.01% solution of humic acid fractions from Haapsalu (Baltic Sea) estuarine mud in 0.9% NaCl solution. Humisol is used (intramuscularly or via electrophoresis) in the cases of chronic radiculites, plexitises, neuralgia, rheumatoid arthritis, arthroses, chronic diseases of tympanum, paranasal sinuses, rhinitis, and other diseases for stimulation of immunity. For the patients treated with Humisol as adjuvant drug, the salmonellosis course was more favorable, the period required for recovery was shorter, and immunity indices earlier became normal (Schepetkin *et al.*, 2002).

Peloidin

Peloidin is a filtrate of specific kind of mud solution from Odessa (Black Sea) estuarine sapropel. Oral administration and electrophoresis of Peloidin are beneficial in lesions of gastric and duodenal mucosa and diseases of the gallbladder and biliary tract (Schepetkin *et al.*, 2002). The drug is also applicable for treating patients with inflammatory processes of the genital system. Peloidin (with phonophoresis) and Humisol (as intramuscular injections) have been demonstrated to cause a distinct increase of cellular immunity indices, and a positive trend in biochemical parameters and cardio-vascular function in patients with pulmonary tuberculosis (Schepetkin, *et al.*, 2002).

Eplir

Eplir is a 1% oil solution of lipid fraction from specific sulfide mud that has anti-oxidant properties. Eplir administration to rats with carbon tetrachloride induced hepatitis protects the liver parench-yma against dystrophy, necrosis, and inflammation (Schepetkin *et al.*, 2002).

FiBS

FiBS (abbreviation of author names: Filatov VP, Biver VA, Skorodinskaya VV) is a product obtained by distillation of specific kind of sea mud. It contains cinnamic acid and coumarins. FiBS possesses immunomodulating action on primary humoral immune response and does not increase a delayed-type hypersensitivity reaction (Schepetkin *et al.*, 2002).

Humex K

Humex K is a natural formulation developed by Unique Formulations in South Africa which can be described as a scientifically proven anti-inflammatory, as well as immunomodulatory and anti-viral. It contains a high strength humic acid which has been long known to be beneficial in relieving inflammatory joints conditions. Clinical trials carried out in association with the University of Pretoria has scientifically established that the use of humic acid is safe for the use of inhibiting inflammatory reactions. Evidence is also available that humic acid can be used as a modulator of inflammation of joints and airways, where it holds promise in autoimmune related diseases and allergic reactions such as osteo-arthritis and hay fever (Humex K, 2008).

Humifulvate and Humet-R

Humifulvate is a standardized complex of peat-derived humic acid, fulvic acid and phenolic acid, intended for human consumption developed by Humet Corporation, Hungary. It is the basic compound that can be used as a dietary supplement following supplementation with minerals and trace elements. The natural source of humifulvate is a unique peat bog in Hungary, on the Northern shore of Lake Balaton estimated to be between 3,000 and 10,000 years old, nourished by the water of a completely unique nearby spring. The chemical and biological properties of this peat bog have been studied for more than 40 years. This wide-scale scientific research has found that the peat deposit is mildly alkaline (pH 7-8) and contains two predominate humate compounds in high quantities: humic acids and fulvic acids, and in a smaller amount phenolic acids. The peat deposit also contains calcium huminate, the degradation product of lignans, shell remnants, calcareous materials, sand and other minerals. Humifulvate is processed into a concentrate for inclusion in dietary supplement products for oral consumption, either in liquid or in solid form.

Humifulvate is a negatively charged metal complexing ligand. There are a number of active sites where metal ions may bind to aromatic and aliphatic carboxyl and phenolic hydroxyl groups within the humifulvate complex, allowing humifulvate to act as an ion exchanger, releasing metal ions of low atomic mass and chelating heavier metals. Research results suggest that Humifulvate is able to prevent mineral toxicity by chelating poisonous metals; such as Cadmium, Lead, Mercury, and Aluminium along the intestinal tract, thereby reducing heavy metal burdens (Humifulvate, 2008).

Humifulvate for human consumption is marketed as Humet-R which is available both as a syrup and as convenient easy-to-take capsules. Humet-R, now known for efficient Heavy Metal Detox, was originally designed to be a mineral supplement that efficiently delivers minerals to the parts of the body that requires them. These minerals are carried in a chelate with Humifulvate. When Humifulvate chelates the minerals, it carries the minerals into the body in the form of organic minerals. This makes the minerals highly bio-available and the body can absorb the minerals very easily. When the Humifulvate has delivered its mineral, rather than then deteriorating and disappearing, it naturally attaches itself to the toxic heavy metals and removes them from the body via the gut (Humet-R, 2008).

3

SHILAJIT

A humic substance with proven therapeutic properties

There are a number of natural remedies which have been in use for ages as part of traditional medicine worldwide, but unfortunately lack systematic scientific evaluation and documentation. The world today is looking at these remedies for a number of ailments. However, these remedies can only find a place for themselves in the mainstream medicine if their claims are evaluated scientifically and documented systematically.

Shilajit is one such remedy, which has been in use as a folk medicine for over 3000 years as a rejuvenator and adaptogen (Sharma, 1978). It has been used by *Vaidyas* and *Hakims* for ages and has a unique place in the ancient Ayurvedic texts. It has been said that there is hardly any curable disease, which cannot be controlled or cured with the aid of shilajit. Although this is a tall order, scientific studies over the last 20–25 years have shown that it is indeed a panacea of traditional medicine, effective in a number of ailments. We present here a brief review of the ancient claims for this panacea and the modern scientific findings, which have validated these claims.

CHEMICAL NATURE OF SHILAJIT

The extensive therapeutic value attributed to shilajit attracted a large number of European, Indian, and Russian scientists to determine its source, chemical nature, and origin. Considerable controversy existed in the literature, till about 30 years back, regarding the exact chemical nature of shilajit when research carried out by Ghosal and coworkers at Banaras Hindu University, India, proved beyond doubt that shilajit was a humic substance. Till then, shilajit was variously described, as an inorganic material, a bitumen, an asphalt, a resin, a mineral, a plant fossil exposed by the elevation of the Himalayas, a substance of mixed plant and animal origin (Ghosal, 1992). Earlier work carried out by Stevenson in 1833, Campbell and Shermill in 1846, Lawder in 1871, Oldham and Leigh in 1921 regarded shilajit as an impure sulfate of alumina. Later on Campbell, Lawder, and Trail regarded shilajit as a bitumen or mineral resin (Lal *et al.*, 1988). Rajnath and Prasad (1942) carried out geological studies and revealed that in spite of exudation of shilajit from rocks, neither a bitumen or mineral resin nor rocks play any major role in it formation. A cactus like plant Thuar (a kind of Euphorbia) along with other vegetations was reported to be growing in the vicinity of shilajit exuding rocks.

Chopra (1958) suggested that shilajit may be a vegetable product but on account of the presence of albuminoids and hippuric acid in it, the possiblility of role of animals could not be ruled out. Singh and Sharma (1970) gave strong geological support to prove that shilajit was not a bitumen and also suggested that some latex or gum-bearing plant may be its origin. Lal and Joshi (1976) analyzed the shilajit bearing rocks collected from the Himalayas and reported that the minerological composition of the rocks were entirely different from the organic nature of the exudate (shilajit), thus ruling out its mineral origin.

The strongest scientific evidence regarding the organic nature and origin of shilajit was provided by Ghosal *et al.* (1976). The analysis of latex of *Euphorbia royleana* collected in summer month from the plants growing in the vicinity of shilajit exuding rocks in Himalayas, revealed the presence of identical organic compounds in it, which provided a strong evidence to prove that the chemical constituents of shilajit are primarily derived from the latex of *Euphorbia royleana*. Another research claimed that mosses like species of *Barbula*, *Fissidenc*, *Minium*, *Thuidium*, and species of Liverworts, which are present in the vicinity of shilajit exuding rocks may be responsible for the formation of shilajit (Joshi *et al.*, 1994). The bryophytes revealed occurrance of minerals and metals in their tissue which were similar to the elements present in shilajit. Later research by Ghosal and co-workers proved that shilajit was mainly composed of fresh and modified remnants of rock humus admixed with rock minerals and other organic substances occurring in the shilajit-bearing rock rhizospheres (Ghosal *et al.*, 1988; Ghosal, 1990, 2006). Humic substances chemically are heteropolycondensates of both organic and inorganic compounds. In sedimentary rocks, the organic compounds derived mainly from lithophilic vegetation (*e.g.*, lichens, mosses, algae, and microorganisms) and marine fossil fauna, get into complex combination with metal oxides and salts and rock minerals by biochemical (*e.g.*, isoenzymes from microorganisms) and geochemical processes to produce humic substances (Ghosal, 2006).

Earlier research on the chemical nature of shilajit showed that its major organic constituents included benzoic acid, hippuric acid, fatty acids, resin and waxy materials, gums, albuminoids, and vegetable matters admixed with some inorganic minerals that produced oxides of Fe, Al, Ca, Mg, K, and P on incineration of shilajit (Chopra *et al.*, 1958; Kong *et al.*, 1987). These constituents were, however, individually or in combination, unable to account for a number of therapeutic properties of shilajit. Despite a lack of evidence, earlier research attributed the therapeutic properties of shilajit essentially to benzoic acid and metallo-benzoates contained in it (Chopra *et al.*, 1958; Kong *et al.*, 1987). Extensive research in the 1980s showed that the major organic mass of shilajit comprised of humus (60–80%) along with other components such as benzoic acid, hippuric acid, fatty acid, ichthyol, ellagic acid, resin, triterpenes, sterol, aromatic carboxylic acid, 3,4-benzocoumarins, amino acids, and phenolic lipids (Ghosal *et al.*, 1988a).

The biological activity of shilajit can be ascribed to three distinct classes of compounds (Ghosal *et al.*, 1991, 2002; Ghosal, 2006):

1. the low-and medium-molecular weight non-humic organic compounds such as oxygenated dibenzo-α-pyrones, both mono- and bis-compounds thereof, in free and metal–ion conjugated forms;

2. medium-and high-molecular-weight dibenzo-α-pyrones chromoproteins, containing trace metal ions and coloring matter, *e.g.*, carotenoids and indigoids; and

3. the medium-molecular-weight humic and fulvic acids which act as carriers for the bioactive principles and help in their *in vivo* transportation in the body.

The composition of shilajit is influenced by factors such as the plant-species involved, the geological nature of the rock, local temperature profiles, humidity, and altitude, etc. For example, it was found that shilajit obtained from India in the region of Kumoan contains higher percentage of fulvic acids (21.4%) as compared to shilajit obtained from Nepal (15.4%), Pakistan (15.5%), and Russia (19.0%). On the other hand, the bioactive low-molecular compound was found to be high in shilajit obtained from Nepal. Similarly, pH of the 1% aqueous solution of shilajit was different for samples obtained from different countries, *viz.*, 6.2 for India (Kumoan), 7.5 for Nepal (Dolpa), 6.8 for Pakistan (Peshawar), and 8.2 for Russia (Tien-Shan). Similarly, humic constituents in shilajit samples obtained from these countries also varied (Ghosal *et al.*, 1991b). This may be due to the fact that humus reserve is a complex mass whose complexity is determined by the intensity of several factors such as rate of formation of fresh humus, adsorption of plant exudates, debris and microorganisms to humus reserve, the rate of decomposition of the free and entrapped low-molecular-weight. components and the rate of decomposition of HAs and FAs into humin and other intractable products.

NONHUMIC ORGANIC COMPONENTS OF SHILAJIT

The nonhumic organic compounds of shilajit comprise of low-and medium-molecular-weight compounds (Figure 3.1) (Ghosal, 1989, 1990, 2002, 2006), mainly the oxygenated dibenzo-α-pyrones both in conjugated and nonconjugated forms. Some of these components isolated from shilajit by Ghosal and coworkers included 3,8-dihydroxydibenzo-α-pyrone, 3-hydroxydibenzo-α-pyrone, dimeric and tetrameric dibenzo-α-pyrone and their hemiquinones, and hydroaromatic dibenzo-α-pyrone. In shilajit, dibenzo-α-pyrones and their oligomeric equivalents occur to the extent of 0.1–2%, in the free state while a larger relative proportion (10–30%) is present in conjugate form. These components are the central chemical characters of shilajit organic matter and constitute the core nucleus of shilajit (rock) humuspaleohumus. Dibenzo-α-pyrones are capable of permeating through the blood-brain barrier and act as a powerful antioxidant, protecting the brain and nerve tissue from free-radical damage (Schepetkin *et al.*, 2002).

(1)
Where R^1 is selected from the group consisting of H, OH, O-acyl and O-amino-acyl; and R5, R6, R7, R8, R9, and R10 are independently selected from the group consisting of H, OH, O-acyl, O-amino-acyl, and fatty acyl groups

Dimeric DBP & its Hemiquinone form

Tetreameric DBP & its Hemiquinone form

C_{20} : 5 ω 3, Eicosapentaenoic acid
EPA

C_{22} : 5 ω 3, Docosapentaenoic acid
DPA

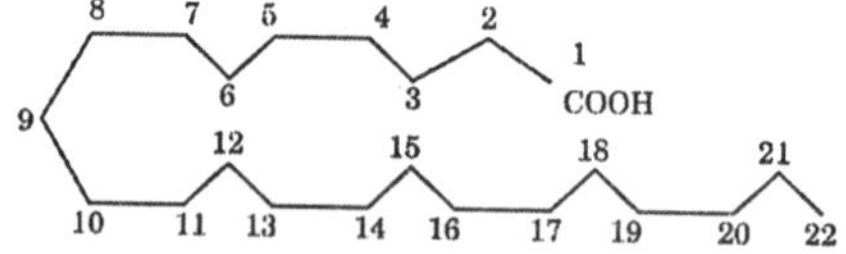

C_{22} : 6 ω 3, Docosahexaenoic acid
DHA

EPA : ω-3-polysaturated fatty acids; (C_{20} : 5 ω 3, eicosapentaenoic acid)
DPA : ω-3-polysaturated fatty acids
DHA : ω-3-polysaturated fatty acids; (C_{22} : 6 ω 3, eicosapentaenoic acid)

Fig. 3.1: Low-and medium-molecular-weight components of shilajit (Ghosal, 2002)

OXYGENATED DIBENZO-α-PYRONE CHROMOPROTEINS (DCPS)

An important class of compounds, namely, oxygenated dibenzo-α-pyrone chromoproteins have been isolated and identified from shilajit (Ghosal, 2005, 2006). Dibenzo-α-pyrone chromoproteins are aggregates and complexes of oxygenated dibenzo-α-pyrones with metal ions, chromo constituents, apo-proteins and lipids present as lipoproteins. Figure 3.2 gives the structure of DCPs, while figure 3.3 shows a schematic representation of the constituent assembly of DCPs. These special class of dibenzo-α-pyrone conjugtes occur to the extent of 10–18% in the water-soluble fraction of shilajt and constitute its principal bioactive substances.

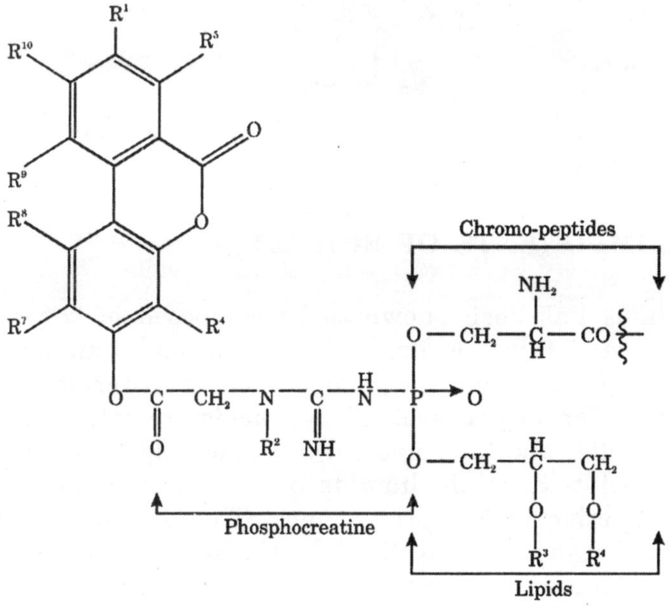

Where:

R^1 : H, OH, O-acyl, O-amino acyl, or di- or tri-peptides of these aminoacids;
R^2 : H or CH3;
R^3 : H or C14-C24 saturated or unsaturated fatty acid; degree of unsaturation ranging from 1 to 6; and
R^5, R^6, R^7, R^8, R^9, and R^{10} are independently selected from the group consisting of H, OH, O-acyl, O-amino-acyl and fatty acyl groups.

Fig. 3.2: Dibenzo-α-pyrone chromoproteins (Ghosal, 2005)

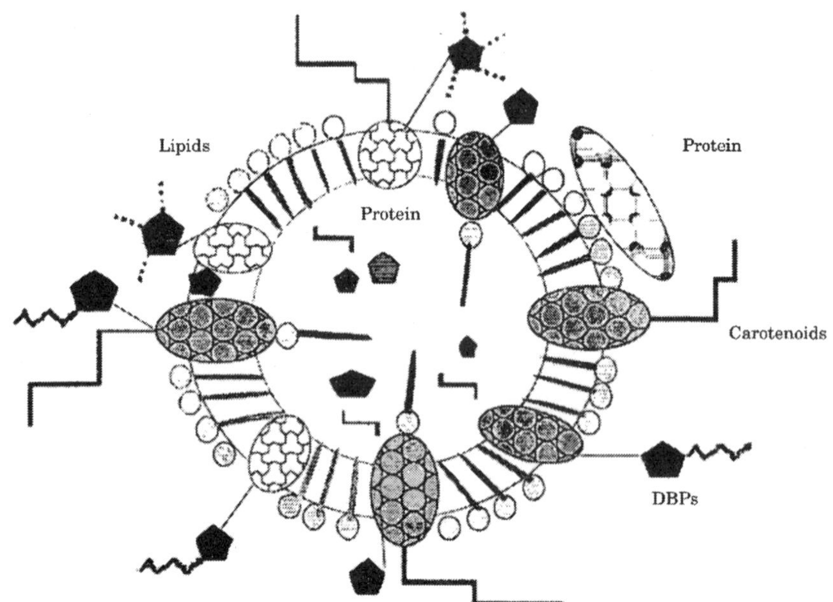

Fig. 3.3: Dibenzo-α-pyrone chromoproteins constituent assembly (Ghosal, 2005)

HUMIC COMPONENTS OF SHILAJIT

Shilajit humus has been shown to be composed of three major components, *viz.*, fulvic acids, humic acids, and humins, which can be distinguished on the basis of their different aqueous solubility at different pH levels. Fulvic acids are soluble in water at all pH (acidic, alkaline, neutral). Humic acids are soluble in water only pH > 3, while humins are practically insoluble in water, irrespective of the pH changes. This solubility property was humic substances was utilized for the separation of the three humus components of shilajit (Ghosal, 1988, 2006).

Fulvic acids have lower molecular weights (0.5–2 kDa) and a smaller number of total and aromatic carbons than humic acids (2–5 kDa), which in turn have longer-chain fatty acid fragments and therefore possess higher hydrophobicity than fulvic acids. Humic and fulvic acids contain carboxyl substituents in aromatic rings. Their aromatic nuclei have a low degree of condensation and are alternated with parts that are nonaromatic. The presence of conjugated *p*-electrons in aromatic rings and various functional groups as substituents in combination with the centers of paramagnetic character allows the substances to form complexes, to participate in ionic exchange and oxidation–reduction processes, to react in numerous tautomeric forms the properties important for the biological action of these acids (Schepetkin *et al.*, 2002).

The humic substances from shilajit, although exhibited a number of similarities in properties with those of soil humus (Schnitzer, 1978), also had certain distinctive features, not found in soil humus (Ghosal, 1993). While oxygenated dibenzo-α-pyrones constituted the core building blocks of rock-humus, simple phenols and phenolic acids from plants constituted the primary building blocks of soil humus. Also, the role of rock-rhizospheric minerals is very important in the formation and stability of rock-humus and the nonhumic metabolites in shilajit. In contrast, soil humus is not endowed with such rich sources of minerals and their attendant attributes (Ghosal, 2006).

The humic substances in shilajit, *viz*., humic acids and fulvic acids have an "open" flexible structure perforated by voids of varying dimensions. These voids are capable of entrapping bioactive molecules like the low-molecular-weight dibenzo-α-pyrones present in shilajit. Such entrapment affords a high degree of protection and stability to the entrapped molecules in their natural habitat. It has also been suggested that humic and fulvic acids act as carrier molecules for delivering these bioactive molecules at their intended site of action.

Figure 3.4 shows a proposed structure of fulvic acids from shilajit and Figure 3.5 shows a schematic diagram for the formation of fulvic acids in shilajit.

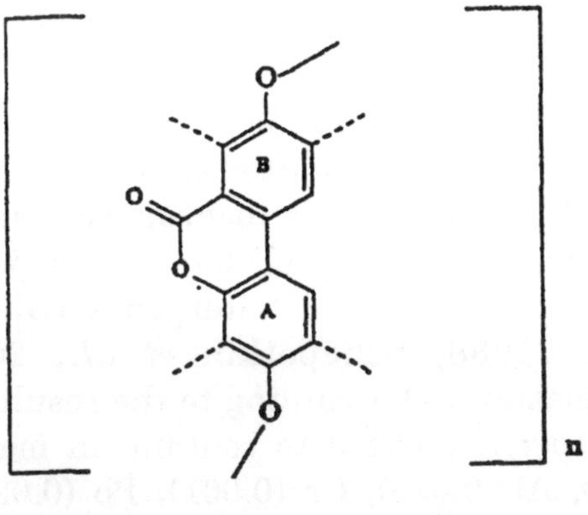

Fig. 3.4: Fulvic acids in shilajit (Ghosal, 2002)

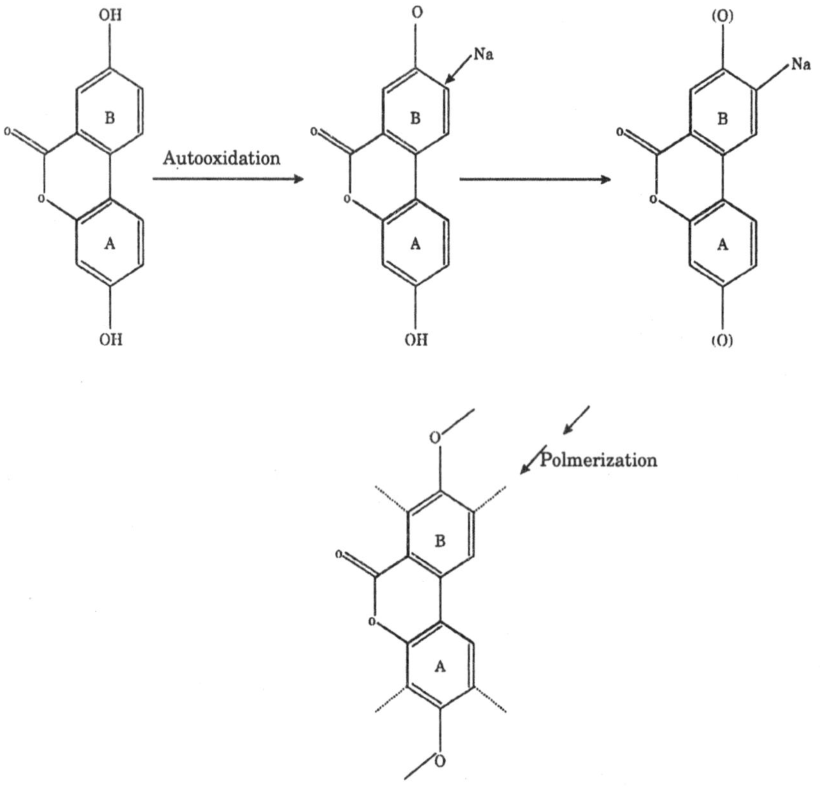

Where Nu is a nucleophile *e.g.*, RO', RNH⁺, RCO2⁺ etc

Fig. 3.5: Schematic representation for the formation of fulvic acids from 3,8 dihydroxy dibenzo-α-pyrones in shilajit (Ghosal, 2003)

INORGANIC AND OTHER MINOR COMPONENTS OF SHILAJIT

Besides the humic and nonhumic organic components discussed above, shilajit has also been reported to contain albuminoids, amino acids (0.23–0.25%), organic acids (benzoic acid and its derivatives, hippuric acid, naphthenic acids), phenolic lipids, polymeric quinones, sterols, tannins, terpenes, and triterpenes (Ghosal *et al.*, 1976; Acharya *et al.*, 1988; Schepetkin *et al.*, 2002). Shilajit is organomineral matter and according to the results of microelement analysis, it has been reported to contain (in mg%): Cu (0.02), Zn (0.01), Li (20.0), Al (0.025), Cr (0.001), Pb (0.02), Ag (0.001), Co (0.01), Hg (0.002), P (0.008), Cd (0.05), Br (0.03), V (0.0016), Fe

(0.16); Ca (31–39 mM/L), mM/L) Mg (7.5–10 mM/L), K (100.0–106.6 mEq/L), As, Na, Cl, I, Mn, Mo, S, Si (Schepetkin *et al.*, 2002). Shilajit also contains ellagic and tannic acids, which are natural polyphenolic antioxidants. Ali *et al.* (2004) has recently reported the isolation of five new phenolic constituents, namely, shilajitol, shilacatechol, shilaxanthone-shilanthranil, naphsilajitone along with a geranyl-acetate–type monoterpene, shilajityl acetate, and a pyrocatechol (Figure 3.6).

ODOR OF SHILAJIT

Two odorous varieties of shilajit have been reported in Ayurveda: *Gomutra* shilajit (smell – like cow urine) and *Karpurgandha* shilajit (smell – like camphor) (Ghosal, 1994a). The chemical constituents of the odor are also believed to contribute to the healing and rejuvenating properties (rasayan) of shilajit. The consistency with which the typical odor of shilajit is associated in widely divergent and distant mountain habitats is indeed remarkable. Despite exposure for ages to relentless onslaughts (extreme weather conditions, cosmic radiation, microbial infestation) in the natural habitats this typical odor of shilajit persisted (Ghosal, 2006). Scientific investigations using head space gas chromatographic technique revealed that the odor of Gomutra shilajit was due to the presence of a number of constituents including ethyl hexanoic acid, *m*- and *p*-cresol, 3,4-ethyl phenol, naphthalene, benzothiazoles, 2,4-dimethylquinoline, and 2,4-bis (1,1-dimethylethyl) phenol. Among the substances responsible for fixing the odor (fixers) of shilajit, high-molecular-weight aliphatic hydrocarbons, the corresponding alkanals and alkanols, triglycerides, 2,4-dimethyl quinoline, benzothiazoles, and phospholipids (containing saturated and poly-unsaturated fatty acyl and O-alkylether moieties) were prominent (Ghosal, 1994b, 2006; Ghosal *et al.*, 1995b).

The typical camphor-like smell of karpurgandha shilajit was attributed to the presence of *p*-cymene and equivalents, *e.g.* isopropyl and methylpropylbenzenes. When present in appreciable amounts, these compounds imparted camphor-like smell to shilajit. The perception and intensity of these odors were found to be dependent on the occurrence and relative abundance of the mentioned odorous constituents in shilajit in different natural habitats, particularly those where the impact of geothermal transformations was pronounced (Ghosal, 1994b, 2006).

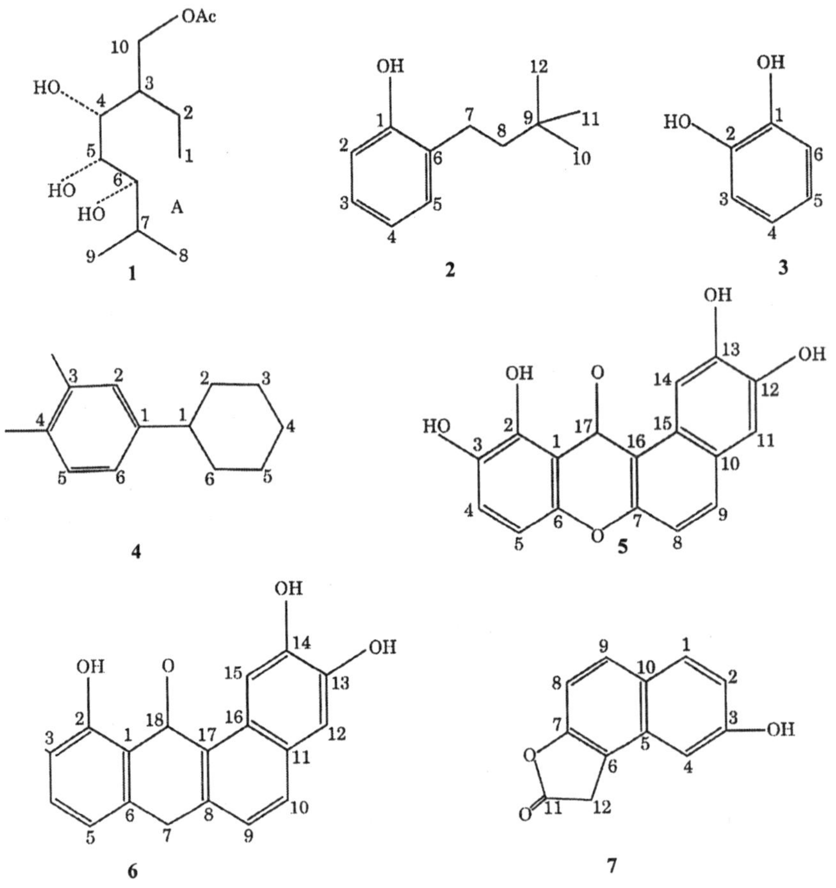

Fig. 3.6: Phenolic components from shilajit (Ali *et al.*, 2004)

PHARMACOLOGICAL PROPERTIES OF SHILAJIT

Although the therapeutic benefits of shilajit have been known for thousands of years, the reason or mechanism of these effects was still not very clear. It is only in the last 50 years that a number of pharmacological properties of shilajit have been proven by systematic experimentation in animals, and the reasons behind these properties have been deciphered.

Antiulcerogenic and Anti-inflammatory Activity

Shilajit is perhaps the only agent to possess both antiulcerogenic and anti-inflammatory activities in a single compound and this unique property of shilajit can be safely utilized in clinical practice (Goel *et al.*, 1990). Shilajit was found to have potent anti-inflammatory activity in all the three models of acute, subacute, and chronic inflammation. Shilajit, at a dose of 50 mg/kg was found to significantly reduce carrageenan-induced hind paw edema in rats having an effect comparable to phenylbutazone (100 mg/kg, i.p.) and betamethasone (0.25 mg/kg, i.p.) (Goel *et al.*, 1990). Studies were carried out on shilajit samples collected from different locations to evaluate their possible role as an antiulcerogenic and anti-inflammatory agents. It was found that shilajit increased the carbohydrate/protein ratio and decreased the gastric ulcer index, indicating an increased mucus barrier (Ghosal *et al.*, 1988b). These results substantiate the use of shilajit in peptic ulcer (Goel, 1990). Isolated constituents of shilajit, *viz.*, fulvic acid containing di-benzopyrones and 4'-methoxy-6-carbomethoxy bi phenyl were also found to significantly reduce the restraint-stress-induced ulcer index in pylorus-ligated albino rats, compared to control and aspirin-treated groups, demonstrating an ulcer protective effect (Ghosal, 1988). The effect was mainly attributed to a decrease in the volume of gastric secretion as well as acid and peptic output and an increase in the mucosal secretion.

In Russia, mumie extract has been found to be highly effective during treatment-for paradontosis in humans and has significant anti-inflammatory effect on osteoarthrosis, rheumatoid arthritis, ankylosing spondylitis, and cervical spondylosis (Schepetkin *et al.*, 2002).

Antioxidant Activity

The uncontrolled production of oxygen-, sulfur- and nitrogen-centered free radicals have been implicated in a number of diseases and debility conditions in humans ranging from arthritis and hemorrhagic shock to AIDS, Alzheimer's disease, and aging. Agents that can regulate the uncontrolled systemic production of these biogenic free radicals can presumably provide cellular protection and regress cellular damage. Processed shilajit has been shown to possess such radical scavanging and antioxidant effect against these biogenic free radicals, thus validating its claim as an important rasayana (revitalizer) (Bhattacharya *et al.*, 1995;

Ghosal *et al.*, 1995d). In a study by Ghosal and Bhattacharya (1996), the antioxidant property of processed shilajit was compared to unprocessed shilajit and vitamin C (ascorbic acid). Processed shilajit not only exhibited a significant antioxidant activity of itself but also had the ability to regenerate (recycle) ascorbic acid after it had neutralized free radicals. The dihydroxybenzo-α-pyrones in shilajit caused recycling (regeneration) of ascorbic acid. Unprocessed shilajit did not consistently exhibit the antioxidant activity.

In another experiment, processed shilajit was tested for its ability to neutralize sulfite anion, hydroxyl, and nitric oxide free radicals. Chemical polymerization by free radicals was measured with and without processed shilajit. Processed shilajit provided almost complete protection of methyl methacrylate against hydroxyl-radical-induced polymerization and significantly inhibited the polymerization of methylmethacrylate by the sulfite free radical. Processed shilajit efficiently trapped nitric oxide free radicals. The antioxidant effects were concentration dependent. Higher concentrations of processed shilajit provided greater free radical protection (Ghosal *et al.,* 1995a; Bhattacharya *et al.,* 1995).

In a separate experiment, the effect of shilajit on lipid peroxidation and gluthathione content in rat liver homogenates was also investigated. It was found that shilajit inhibited lipid peroxidation induced by cumene hydroperoxide and ADP/Fe^{2+} complex in a dose-dependent manner (Ghosal, 2000). Shilajit also decreased the rate of oxidation of reduced glutathione content and inhibited the ongoing lipid peroxidation which was induced by these agents immediately after its addition to the incubation system (Tripathi *et al.,* 1996).

Antidiabetic Activity

Shilajit at a dose of 100 mg/kg, PO was found to decrease streptozocin (STZ)-induced hyperglycaemia in rats. It also reduced the STZ–induced decrease in superoxide dismutase activity in pancreatic islet cells (Bhattacharya, 1995; Kanikkannan *et al.*, 1995). Importantly, it had no effect on the blood glucose level in normal rats. Diabetes mellitus was produced in male albino rats by the administration of STZ 45 mg/kg S.C. on two consecutive days. Hyperglycemia along with superoxide dismutase activity of pancreatic islet cells was assessed on day 7, 14, 21, and 28 days following STZ administration. In separate two other groups, shilajit at a dose of 50 and 100 mg/kg, PO was administered concurrently from 28 days. It was found that STZ-induced significant hyperglycemia by day 14, which was further, increased progressively on day 21 and 28. Similarly, STZ also induced a decrease in pancreatic islet cell superoxide dismutase activity which was apparent on day 7 and increased progressively, thereafter on day 14, 21, and 28. Shilajit at a dose of 50 and 100 mg/kg, PO had no dispersible per se effect on blood glucose level in normal rats but attenuated the hyperglycemic response of STZ from day 14 onwards, though only the effect of the higher dose was statistically significant. Similarly, both the doses, *i.e.*, 50 and 100 mg/kg PO of shilajit reduced the STZ–induced decrease in superoxide dismutase activity from day 14 onwards, the effect of lower dose being statistically insignificant. Earlier observation that STZ–induced hyperglycemia may be due to decrease in pancreatic islet superoxide dismutase activity, leading to accumulation of free radical and damage of β-cells has been confirmed by these experiments. Shilajit prevents both the effect of STZ possibly by its action as a free radical scavenger. This experiment supports the earlier writing of Ayurveda that shilajit can prevent maturity onset diabetes mellitus (Bhattacharya, 1995).

Antistress activity

Shilajit collected from India, Nepal, Pakistan, and Russia and organic constituents isolated from them were studied for their antistress effect in albino mice. It was found that shilajit from Kumoan (India), Dolpa (Nepal), and a combination of the total ethyl acetate extract and fulvic acids extracted from Kumoan shilajit produced statistically significant improvement in forced-swimming-induced immobility in albino mice (Ghosal *et al.*, 1991).

Nootropic (memory enhancement) and Anxiolytic Activity

Nootropics are a class of drugs which act as cognitive and learning enhancers. It has been proposed that the modern equivalent of a medha rasayanas are those substances with nootropic activity (Ghosal, 1998). Medhya is defined as causing or generating intelligence, mental vigor or power.

Studies were carried out to test the validity of use of shilajit as an Ayurvedic medha rasayana (enhancer of memory and learning) in albino rats. Processed shilajit, native shilajit (unprocessed water-soluble fraction), and a preparation consisting of a mixture of ethyl acetate extractive and medium-molecular-weight fulvic acids obtained from processed shilajit were evaluated for putative nootropic and anxiolytic activity in charles foster strain albino rats. The nootropic activity was assessed by passive avoidance learning acquisition and retention while the anxiolytic activity was studied and evaluated by the elevated plus-maze technique. The results of these studies indicated that shilajit had significant nootropic and anxiolytic activity. It was found that processed shilajit and its active constituents (total ethyl acetate fraction and fulvic acids) significantly increased the learning acquisition and memory retention in old albino rats. However, native shilajit produced erratic response (both augmentive and retendative) in the above parameters (Ghosal *et al.,* 1993). The results also indicated that shilajit has significant anxiolytic activity, comparable qualitatively with that induced by diazepam, in doses lower than that required for nootropic activity.

The biochemical studies carried out for level of monoamines indicated that acute treatment with shilajit had insignificant effect on rat brain monoamines and monoamine metabolite levels. However, it was observed that subacute (5 days) dose treatment caused a decrease in 5-hydroxy indole acetic acid concentration and an increase in the level of dopamine, homovallanic acid and 3,4-dihydroxyphenyl acetic acid concentration with insignificant effect on noradrenaline and 3-methoxy-4-hydrophenylethylene glycol levels. The observed neurochemical studies on shilajit indicate a decrease in rat brain 5-hydroxytryptamine turnover, associated with an increase in dopaminergic activity leading to an increase in memory and anxiolytic activity in albino rats (Jaiswal *et al.*, 1992).

Antiallergic Activity

The effect of shilajit and its main active constituents (fulvic acids containg di-benzopyrones, 4'-methoxy-6 carbomethoxybiphenyl and 3,8-dihydroxydibenzo-α-pyrone) were studied in relation to the degranulation and disruption of mast cell against noxious stimuli. Mast cells are the major source of mediators of allergy and anaphylaxis. Shilajit and its active constituents provided statistically significant protection to antigen-induced degranulation of sensitized mast cells, markedly inhibited the antigen-induced spasm of sensitized guinea-pig ileum and prevented mast cell disruption (Ghosal *et al.*, 1989). These findings are consistent with the therapeutic use of shilajit in the treatment of allergic disorders.

Another parameter by which the anti-allergic effect of shilajit and its constituents was determined was on arachidonic acid metabolism. The effect of shilajit on arachidonic acid metabolism was tested in isolated human neutrophils. Shilajit and its constituents inhibited the biosynthesis of the arachidonic acid-lipoxygenase pathway products, *viz.*, leukotriene-B_4, 5 hydroxyeicosatetraenoic acid, 12-hydroxyeicosatetraenoic acid, and also inhibited the biosynthesis of the cyclo-oxygenase product, 12-hydroxyheptadecatrienoic acid in a dose dependent manner. These findings suggest that the inhibition of synthesis of leukotrienes (and equivalents) by shilajit and its constituents is responsible for their therapeutic action in bronchial asthma (Ghosal, 2006).

Immunomodulatory Activity

Shilajit has been used in Ayurveda as a rasayan, which connotes nonspecific host resistance to diseases by augmentation of cellular functions (rejuvination). Shilajit as an immunomodulator agent was studied in mice that were given either shilajit extract or a placebo. The white blood cell activity was studied and monitored prior to and at intervals after receiving the shilajit extract or a placebo. It was found that the white blood cell activity was increased by shilajit extract. Shilajit and its combined constituents elicited and activated, in different degrees murine peritoneal macrophages and activated splenocytes of tumor-bearing animals at early and later stages of tumor growth (Ghosal *et al.*, 1995a).

Since immune activity alters, in concert, with the levels of neurotransmitters, the effects of shilajit were also determined on the levels of brain monoamines in rats. It was found that shilajit at a dose of 25 and 50 mg/kg i.p. for 5 days significantly reduced the level of 5-hydoxy tryptamine and 5-hydroxy indole acetic acid and increased the level of dopamine, noradrenaline, and their metabolites in rat brain (Bhattacharya and Ghosal, 1992). These changes in neurotransmitter levels are similar to those seen in cases of increased humoral (immune) activity and hence validate its use as an Ayurvedic rasayana.

Anti-AIDS Activity

Agents which have the ability to enhance the immune system or temper drug side effects may complement conventional AIDS (Acquired Immuno Deficiency Syndrome) therapies. Drugs that reduce the viral load of the human immuno deficiency virus (HIV) enable the body to restore white blood cell counts, and, as a result, extend the life span of people afflicted with AIDS. Since shilajit is endowed with both immunopotentiating (Ghosal, 1990, 1992 a, b, 1998; Bhaumik *et al.*, 1993) and viral load reducing properties (Ghosal, 2000, 2002a), it has the potential to compliment conventional anti-AIDS therapy. Clinical studies in AIDS patients with a multi-component natural product-formulation, comprising three essential and three supportive ingredients in which shilajit was one of the essential constituent, was conducted. Out of 36 patients enrolled, 22 who received the treatment with the formulation for 6 months showed positive sign of improvement. Their CD4 and CD8 cell counts were increased from 259 ± 119 (CD4) and 733 ± 483 (CD8) to 356 ± 203 and 984 ± 356, respectively. Ten patients who received the treatment for one year, showed distinct improvement in the symptoms and augmentation in the CD4, 516 ± 272; CD8, 1157 ± 428 cell counts (Ghosal, 2006).

SHILAJIT: A *YOGAVAHA*

A remarkable property of shilajit which has been described in ancient texts is that of a *Yogavaha* (Ghosal *et al.*, 1991b, 1995c. *Yogavaha* is an agent which enhances the properties of other drugs. Shilajit is usually soaked in the decoction of one or more of the following plants: *Shoria robusta*, *Bachanania lactifolia*, *Acacia fernesiana*, *Terminalia tomentosa*, *Catechu nigrum*, *Terminalia chebula*, and *Sida cordifolia*, as this is said to increase their efficacy (Nadkarni, 1976).

Although the yogavaha property of shilajit has been described in ancient literature, the reason for the same has not been described. It is possible that this property may be due to the presence of humic substances, mainly humic and fulvic acids, present in shilajit. These humic substances have an "open" flexible structure perforated by voids of varying dimensions which are capable of entrapping the bioactive molecules of shilajit like the low-molecular-weight dibenzo-α-pyrones. Such entrapment affords a high degree of protection and stability to the entrapped molecules in their natural habitat. It has also been suggested that humic and fulvic acids act as carrier molecules for delivering these bioactive molecules at their intended site of action.

Research carried out in the last 10 years by the authors of this book along with coworkers has proven this property of shilajit to be true even with respect to modern drugs. The findings of the studies are presented in subsequent chapters of this book.

4

PHYSICOCHEMICAL AND SPECTRAL CHARACTERIZATION OF SHILAJIT

A number of investigators have characterized shilajit from different regions of the world based on their physicochemical properties and have found distinct similarities in the physical, spectral, and qualitative chemical composition, although the samples differed slightly in their quantitative chemical composition. Samples from different geographical regions were also found to vary in their biological activity which in turn can be attributed to the difference in the levels of their biologically active ingredients.

Frolova and Kiseleva (1996) have reviewed and reported the chemical composition and spectroscopic characteristics of a number of shilajit samples from different regions of Russia and CIS countries. Kiseleva *et al.* (1996) have reported the fatty acid components of a number of Shilajit samples from various geographical regions of the world. Jung *et al.* (2002) reported the 1H NMR and IR spectra of crude shilajit extracted from the mountain regions of Uzbekistan.

The present authors and coworkers (Khanna, 2005; Anwer, 2005; Khanna *et al.*, 2006; Karmarkar, 2007; Agarwal *et al.*, 2008d) characterized a number of shilajit samples from India based on their physicochemical, spectral, and thermal properties.

PHYSICAL CHARACTERISTICS

Raw or rock shilajit in its native form is generally a semi-hard, brownish black to dark, greasy, black resin (Figure 4.1).

Fig. 4.1: Raw shilajit

On dissolving in water, nearly 30–70% of the weight of rock shilajit passes into solution. Undissolved material includes mineral and plant residues in varying quantities depending on purity of the samples used. Purified shilajit available commercially is almost completely soluble in water and very slightly soluble in common organic solvents like chloroform, ethyl acetate, and alcohol in which only the low-molecular-weight nonhumic components of shilajit dissolve. Ghosal *et al.* (1991) has reported a pH range of 3.8–8.2 for 1% aqueous solution of various shilajit samples collected from different regions of the world. Table 4.1 lists down some of the physical properties of typical raw shilajit samples analyzed by the authors.

Commercial shilajit samples are often dried aqueous extracts of raw shilajit and have a blackish, greasy, resinous consistency (Figure 4.2) and a distinctive coniferous smell and bitter taste.

Fig. 4.2: Commercially available purified shilajit from Gurukul Kangri, India

Table 4.1: Physical characteristics of raw shilajit

Characteristic	Raw shilajit
Nature	Brownish black rock which yielded a brownish black powder on pulverization
Color	Brownish black
Odor	Characteristic coniferous
Taste	Characteristic pungent and astringent
Solubility	Partially soluble in water and alkali. Slightly soluble in ethanol and methanol. Almost insoluble in chloroform and ethyl acetate.
pH of 2% aqueous solution	5.52

ELEMENTAL ANALYSIS

Elemental analysis of crude and purified shilajit samples from different sources [Pioneer Enterprises (PE), Natural Remedies (NR), and Gurukul Kangri (GK)] in India was carried out on an elemental analyzer and the results obtained are depicted in Table 4.2.

Table 4.2: Elemental analysis of shilajit samples

Samples	%N	%C	%S	%H	C/N ratio
Raw shilajit	1.13	19.51	0.31	2.51	17.22
Shilajit extract (PE)	3.25	45.19	1.33	6.65	13.87
Shilajit extract (NR)	2.78	34.49	2.22	4.57	12.39
Shudh Shilajit (GK)	0.83	33.96	0.15	4.95	40.89

Ghosal (2006) (Table 4.3) has also reported the elemental composition of seven different samples of shilajit from various regions and the results obtained are shown in Table 4.3.

Table 4.3: Elemental analysis of shilajit samples (Ghosal, 2006)

	Elemental Composition			
	%N	%C	%S	%H
Mean	5.64	36.33	0.86	5.72
Range	5.03 – 5.91	29.0 – 40.8	0.05 – 0.96	5.03 – 5.91

TRACE ELEMENTAL ANALYSIS

Ghosal (1993, 2006) (Table 4.4) has reported the presence of a number of trace elements in the water-soluble fractions of different shilajit samples when analyzed by atomic absorption spectroscopy and high-performance thin layer chromatography.

Table 4.4: Trace elemental composition of shilajit samples (Ghosal, 1993, 2006)

Trace elements	Percentage composition
Aluminium	0.08 – 1.0
Calcium	2.00 – 16.0
Copper	0.005 – 0.01
Iron	0.5 – 1.8
Potassium	0.1 – 0.8
Magnesium	5.0 – 8.0
Manganese	0.008 – 0.01
Molybdenum	0.002 – 0.003
Sodium	0.1 – 0.5
Phosphorous	0.2 – 2.5
Sulfur	0.2 – 1.1
Silicon	0.8 – 1.2
Vanadium	0.002 – 0.01
Tungsten	0.001 – 0.08
Zinc	0.002 – 0.003

UV-VIS Spectroscopy

UV-vis spectra of various shilajit samples from India were obtained on a Shimadzu, 1601 UV/VIS spectrophotometer by dissolving the various shilajit samples in water and recording the spectra in a 1-cm quartz cuvette by scanning from 200 to 800 nm (Khanna, 2005). Since humic substances usually yield uncharacteristic spectra in the UV and visible region, E4/E6 ratio (ratio of the absorbance of the solution at 465 and 665 nm) (Schnitzer, 1972) was determined for the various samples.

As seen in Figure 4.3, the UV-vis spectra of the various samples of shilajit did not exhibit any sharp maxima and the absorbance value decreased with increasing wavelength. A lack of maxima can be expected considering the complicated, multicomponent nature of shilajit with the variety of chromophores of its constituents contributing variable absorptivities. The results are consistent with those obtained for humic-rich soils and humic substances (Schnitzer, 1972). A shoulder was observed at the region around 270 nm, which is generally attributed to the absorbance of the aromatic groupings which are characteristic of humification (Domeizel *et al.*, 2004). Frolova *et al.* (1996) has also reported a pronounced maximum in the region of 280 nm in the UV spectra of mumijo samples from Russia.

Table 4.4 shows the absorbance value of the different samples at 465 and 665 nm and also the E4/E6 ratio of different samples.

Ghosal *et al.* (1991) has reported a E4/E6 ratio ranging from for shilajit samples from different regions of the world.

Table 4.5: E4/E6 ratios of shilajit from different sources

	Raw shilajit	Shilajit (GK)	Shilajit (NR)	Shilajit (PE)
Absorbance at 465 nm (E4)	0.455	0.655	0.532	0.492
Absorbance at 665 nm (E6)	0.145	0.245	0.159	0.157
E4/E6 ratio	3.14	2.67	3.33	3.13

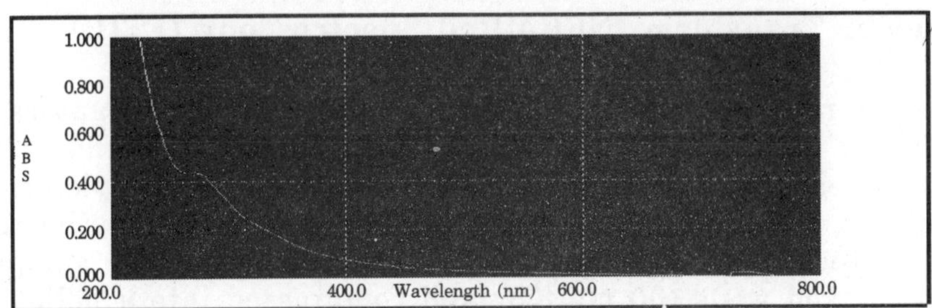

UV–visible spectra of raw shilajit

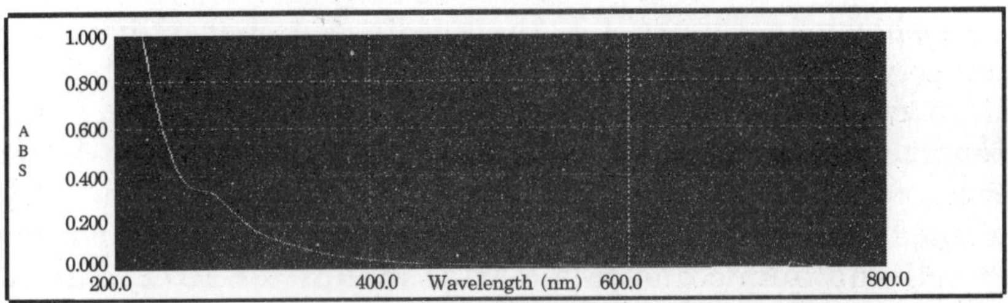

UV–visible spectra of shilajit: Gurukul Kangri

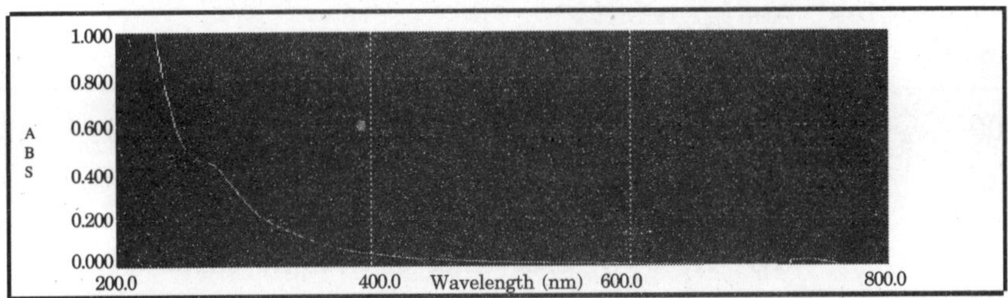

UV–visible spectra of shilajit: Natural Remedies

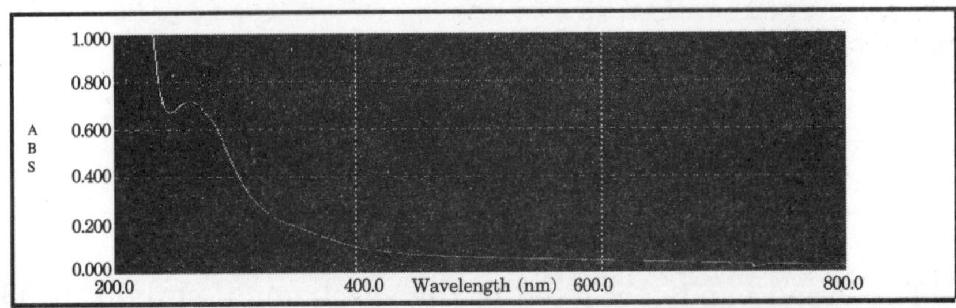

UV-visible spectra of shilajit: Pioneer Enterprises

Fig. 4.3: UV-visible spectra of shilajit from different sources

Fourier Transform INFRA-Red Spectroscopy (FTIR)

IR spectroscopy has been used by a number of investigators as a means of identification of shilajit samples. Khanna *et al.* (2007) recorded the IR spectra of various shilajit samples from India on a FTS 40 (BioRad, USA) FTIR instrument by the KBr pellet technique. Two milligram of previously dried samples were mixed with 100 mg KBr and compressed into a pellet on an IR hydraulic press and the spectra was recorded from 4000 to 450 cm^{-1}.

As shown in Figure 4.4, the specta are characterized by a relatively few bands that are broad. This is expected since shilajit consists of a complex mixture of diverse materials and extensive overlapping of individual absorption peaks because various functional groups present is likely to occur. The results are consistent with those reported in the literature for humus-rich soil and humic and fulvic acids (Schnitzer 1972; Ghosal *et al.*, 1993).

Frequencies (cm^{-1})	**Assignments**
3400	OH stretching, NH stretching.
2900	C-H stretching.
1725	COOH, C=O stretching of ketonic carbonyl
1630	Aromatic C=C, hydrogen bonded C=O of carbonyl, COO-
1450	Aliphatic C-H
1400	COO-, aliphatic C-H
1200	C-O stretch for OH deformation of COOH
1050	SiO of silicate impurities

The broad bands observed in the spectrum of shilajit at 3390 cm^{-1}, 1725 cm^{-1}, and 1638 cm^{-1} can be attributed to hydrogen-bonded OH group, C-O stretching of COOH, and C-C double bond, respectively. The sharp bands observed at 2923 cm^{-1}, 1386 cm^{-1}, and 1036 cm^{-1} can be attributed to the bending vibration of

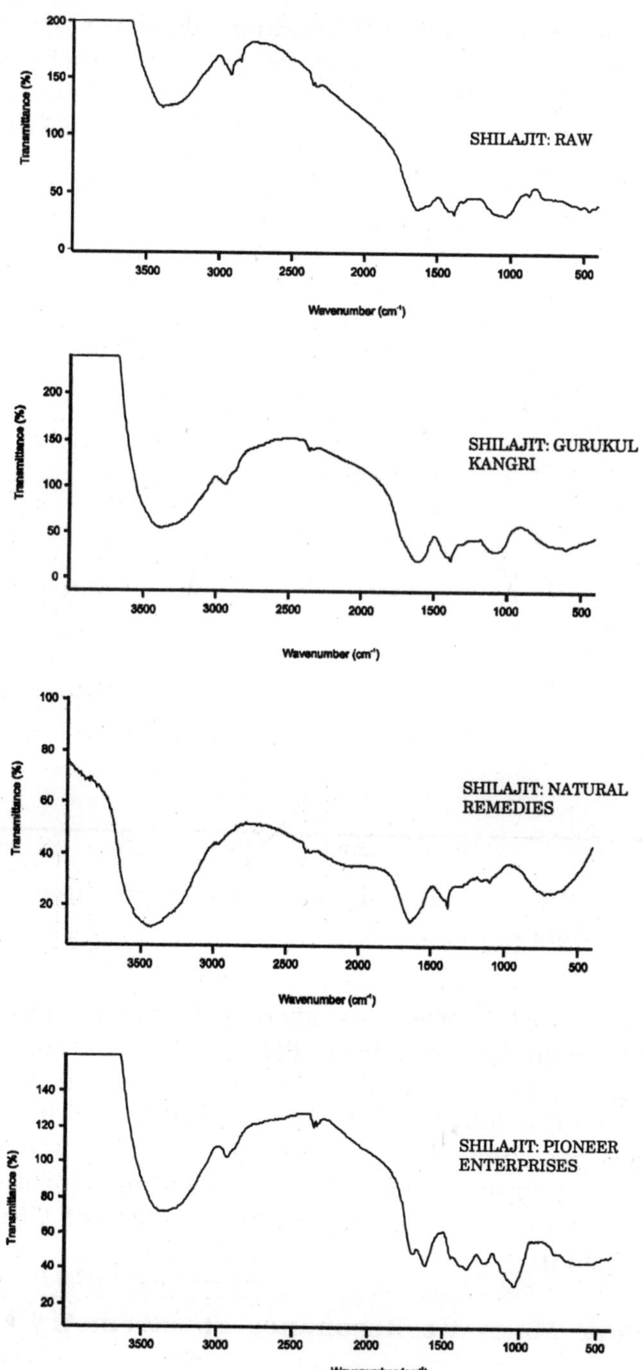

Fig. 4.4: FTIR spectra of shilajit from different sources

aliphatic C-H group, the O-H bending vibrations of alcohols or carboxylic acids, and OH bending deformation of carboxyl groups, respectively. The spectrum reflects the preponderance of oxygen-containing functional groups like COOH, OH, and C-O in shilajit.

Jung *et al.* (2002) has reported the IR spectra of dried aqueous extract of mumie sample collected from the mountain regions of Uzbekistan (Figure 4.5). The spectra exhibited characteristic absorption bands typical for humic substances with major bands seen near 3400 cm^{-1} (H-bonded OH), 2900 cm^{-1} (aliphatic C-H stretching), 1600-1720 cm^{-1} (aromatic C=C, H bonded C=O, and C=O stretching of COOH and ketonic C=O), and 1400 cm^{-1} (C-H bending of CH2 or CH3 groups).

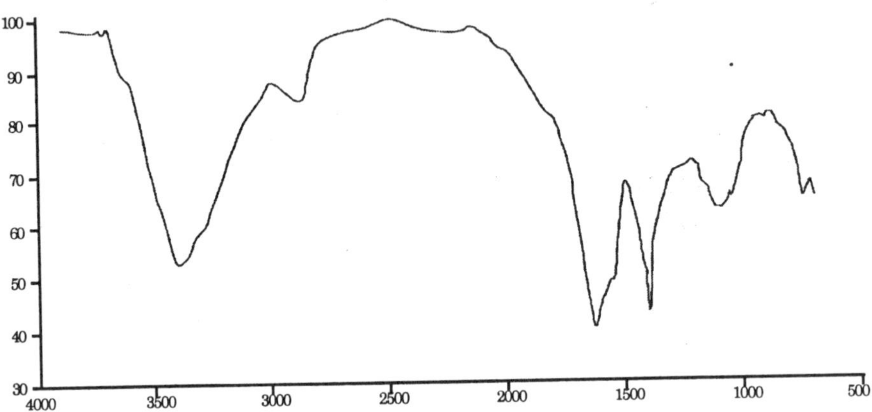

Fig. 4.5: FTIR spectra of dried mumie extract from Uzbekistan (Jung *et al.*, 2002)

Frolova *et al.* (1996) has also reported the IR spectra of various Mumijo samples from Russia. The IR spectra of the samples generally exhibited similar characteristic bands in the region of 3400-3450 cm^{-1} (assigned to the stretching vibrations of OH groups), 2960 cm^{-1} (stretching vibrations of CH3 groups in aliphatic hydrocarbons), 730 cm^{-1} (bending vibrations of the OH group), and 1550 cm^{-1} (bending vibrations of CH2- in the -CH2C0- group).

^{1}H Nuclear Magnetic Resonance Spectrometry

^{1}H NMR spectra of shilajit samples from India were recorded using a Bruker DRX-300 NMR spectrometer. All the spectra were recorded in deuterated dimethylsulfoxide (DMSO-d6) solvent. DMSO-d6 was preferred to other NMR organic solvents since it is a good solvent for humic substances (Ruggiero *et al.*, 1979).

Figure 4.6 shows a typical ^{1}H NMR spectra obtained with shilajit sample from Natural Remedies, India. This and spectra from other samples exhibited a combination of broad and sharp absorption signals which suggest the presence of high-molecular-weight, complex, heterogenic, multicomponent system, as would be expected. The spectra may be divided into three main regions: (1) 0.5–2 ppm, signals due to methyl and methylene protons; (2) 2–5 ppm, signal due to H in structures containing methyl or methylene groups linked to heteroatoms, such as O and N, as well as H of OH groups in alcohols, phenols, and carboxyls; and (3) 6–8 ppm, signals due to aromatic protons. The results are similar to those observed in case of humic substances from other sources (Garcia *et al.*, 1994; Hanninen *et al.*, 1993; Ruggiero *et al.*, 1980).

Jung *et al.* (2002) has reported the ^{1}H NMR spectrum of the crude mumie extract from Uzbekistan on a Bruker DRX500 high-resolution spectrometer operating at 500.13 MHz (Figure 4.7). The sample solution was prepared by dissolving approximately 100 mg of the dry mumie extract in 1 mL of D2O.

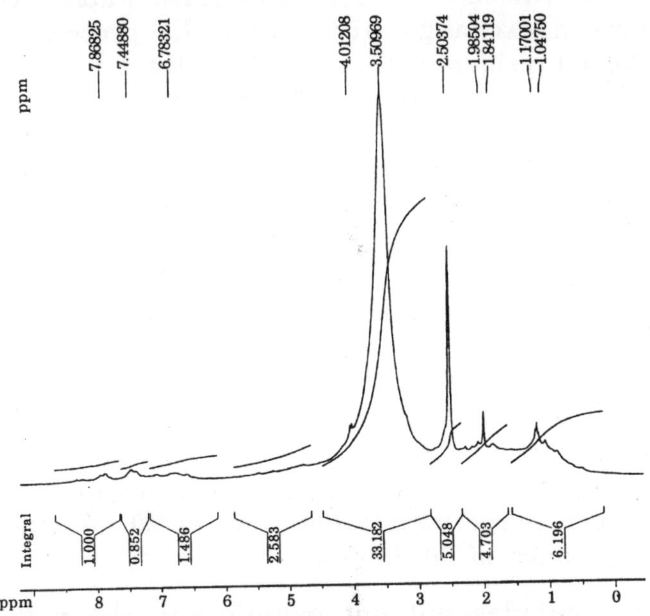

Fig. 4.6: ^{1}H NMR spectra of shilajit sample from Natural Remedies, India

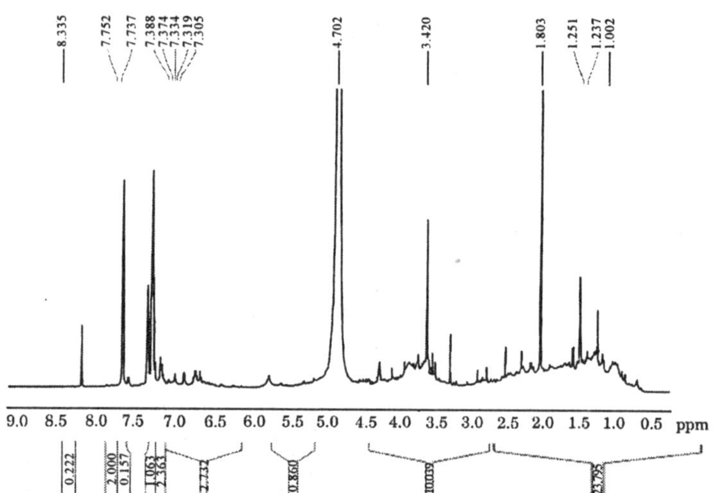

Fig. 4.7: ^{1}H NMR spectra of dried mumie extract from Uzbekistan (Jung *et al.*, 2002)

The spectrum exhibited a combination of narrow and broad signals. The broad peaks at 6.0–8.0 could be assigned to aromatic compounds; peaks at 5.7 and 3.0–4.5 to polysaccharides, and peaks at 0.5–3.0 ppm to aliphatic components. The strong peak at 4.7 ppm could be attributed to monodeuterated water formed from the deuterium exchange with the acidic protons of mumie compounds and residual protons of D2O solvent. The sharp peaks seen in the spectra at 7.3–7.5 ppm (aromatic) can be assigned to benzoic acid or its derivatives, and peaks near 1.8 ppm may be arising from acetic acid. Signal at 1.2 ppm is likely to be caused by isopropanol, and peaks at 1.0 and 0.7 ppm were assigned to oil impurity (Jung *et al.*, 2002).

Differential Scanning Calorimetry (DSC)

DSC thermograms (Figure 4.8) of various shilajit samples from different sources in India were obtained on a Perkin-Elmer Pyris 6 instrument. Samples (2–3 mg) were accurately weighed and heated in closed aluminium crimp cells at a rate of 10 °C/min under nitrogen purge with a flow rate of 20 mL/min over the temperature range of 50–300 °C.

Shilajit samples did not exhibit any sharp endotherm indicating the absence of a well-defined melting point. A shallow

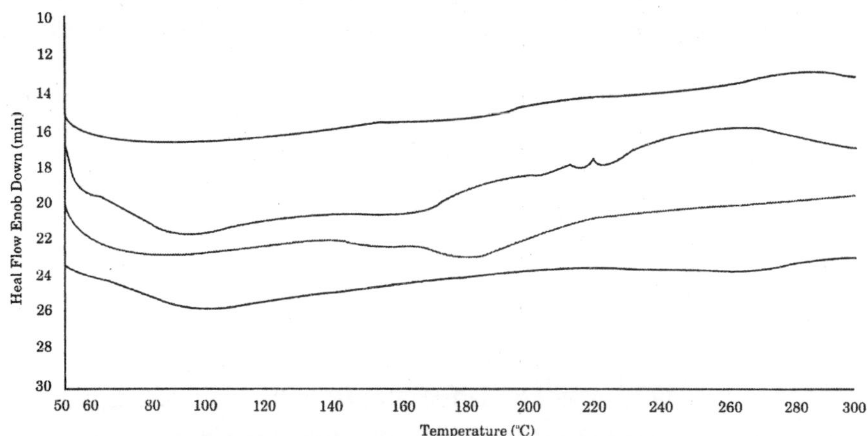

Fig. 4.8: DSC spectra of shilajit from different sources

endotherm could be observed near 90 °C which could be attributed to dehydration of the samples. An exothermic event could be observed at a temperature above 250 °C which could be attributed to the thermal degradation of carbohydrates, dehydration of aliphatic structures, and decarboxylation of carboxylic groups (Pietra and Paola, 2004). A few more thermal events seen in some of the samples could be attributed to the presence of small amounts of lipid crystallites and other components of natural organic matter (Chilom and Rice, 2005).

Powder X-Ray Diffraction

Powder X-ray diffraction patterns of powdered samples of shilajit were obtained using a Panalytical X-ray diffractometer PW3719. All the samples were treated according to the following specifications:

Target/filter (monochromator)	:	Cu
Voltage/current	:	40 kV/50 mA
Scan speed	:	4°/min
Smoothing	:	0

Figure 4.9 shows a typical X-ray diffractogram of raw shilajit, exhibiting a largely noncrystalline nature as evident from the absence of sharp diffraction peaks. A few peaks were, however, observed which could be attributed to the presence of lipid crystallites, clay particles, and other crystalline and micro-

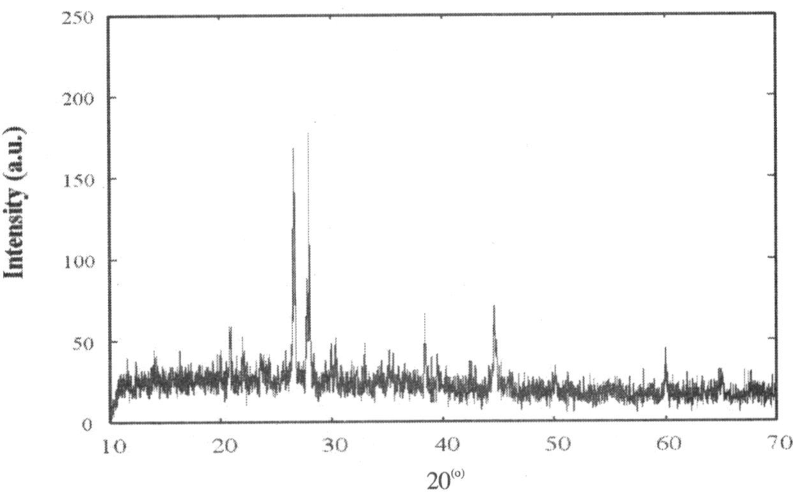

Fig. 4.9: X-ray diffraction pattern of raw shilajit

crystalline materials in the shilajit sample. The behavior is consistent with the multi-component nature of shilajit and with the behavior observed in case of humic substances from other sources (Chilom and Rice, 2005; Visser and Mendel, 1971).

Scanning Electron Microscopy (SEM)

Scanning electron micrographs of powdered shilajit samples were obtained using a Joel JSM-840 Scanning Microscope with a 10

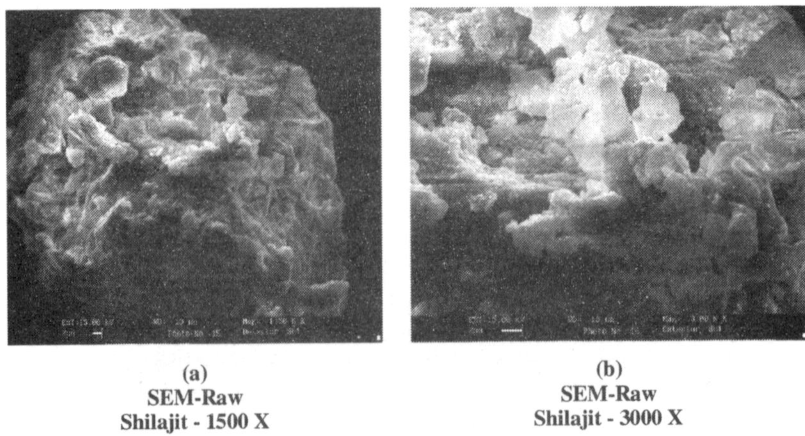

(a) SEM-Raw Shilajit - 1500 X

(b) SEM-Raw Shilajit - 3000 X

Fig. 4.10: Scanning electron micrographs of raw shilajit

kV accelerating voltage. The surfaces of samples for SEM were made electrically conductive in a sputtering apparatus (Fine Coat Ion Sputter JFC-1100) by evaporation of gold.

Figure 4.10 (a,b) shows scanning electron micrographs of a powdered samples of raw shilajit at magnifications of 1500× and 3000×, respectively. The shilajit particles exhibited a spongy structure with the presence of internal spaces. The particles also showed a tendency to aggregate.

5

EXTRACTION AND CHARACTERIZATION OF HUMIC SUBSTANCES FROM SHILAJIT

Humic substances, *viz.*, fulvic acid (FA), humic acid (HA), and humin (HM) have traditionally been separated on the basis of their differential solubility in water at different pH levels. FA is soluble in water at all pH (acidic, alkaline, neutral). HA is soluble in water only at pH > 3, while HM is practically insoluble in water, irrespective of the pH changes. This differential solubility property has also been utilized by a number of researchers for the separation of the three humus components from shilajit.

Ghosal *et al.* (1989) utilized the approach described in Figure 5.1 for the separation of humin, humic, and fulvic acids from shilajit. Since shilajit contains a number of small organic molecules entrapped in the voids of its humic substances, these were first separated by exhaustive extraction of the shilajit-powder with hot ethylacetate. The ethylacetate-insoluble material was treated with cold, very dilute, hydrochloric acid to remove the basic protein conjugates as aqueous acid-soluble fraction. The insoluble solid, comprising the three classes of humic substances was triturated with 0.1 N-NaOH(aq) under an atmosphere of nitrogen, at room temperature for 1–2 hours. The mixture was filtered to get rid of the high-molecular-weight polymeric materials (quinones, toxins, heavy-metal-ion conjugates) and humin as the alkali-insoluble fraction. The filtrate was acidified (pH 3) with HCl and the solution was allowed to stand, at room temperature, overnight when the high-molecular-weight HA got precipitated. The HA were separated by centrifugation and the supernatant, consisting of FA, was passed through activated charcoal to obtain a pale-yellow filtrate. It was evaporated, under reduced pressure, to give a mixture of

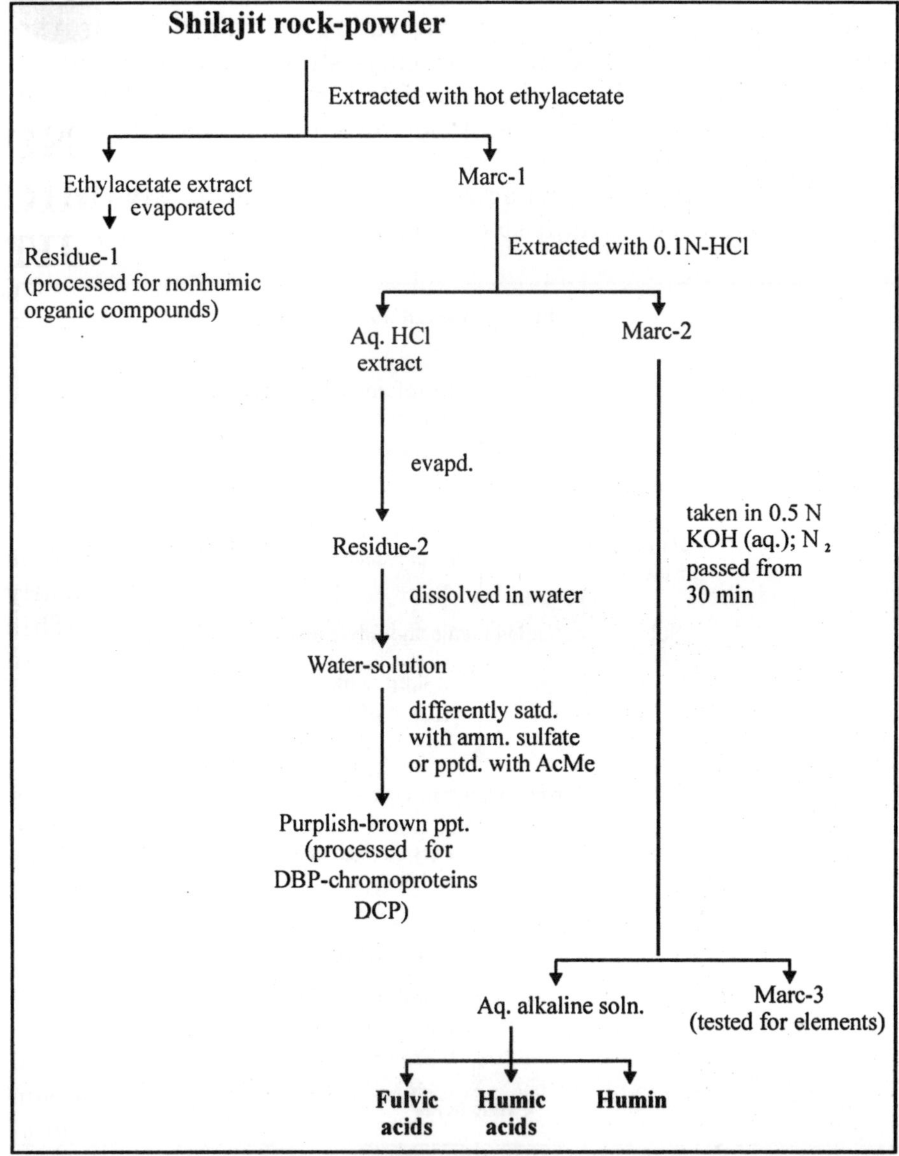

Fig. 5.1: Schematic representation of the method adopted by Ghosal *et al.* (1988) for separation of humic substances from shilajit

low-to-high molecular weight fulvic acids. Another crop of FA that remained adsorbed on activated charcoal was eluted with acetone. The acetone solution, on evaporation, afforded a yellowish-brown powder comprising low-and-medium molecular weight FA (Ghosal *et al.*, 1989).

Khanna (2005) developed an improved method for the extraction of FA from shilajit, which is shown schematically in Figure 5.2. Briefly, the method consisted of successive extraction of raw shilajit with hot organic solvents of increasing polarity (chloroform, ethyl acetate, and methanol) to remove the bioactive components, specifically, oxygenated dibenzo-α-pyrones. Extracted shilajit so obtained was taken and dispersed

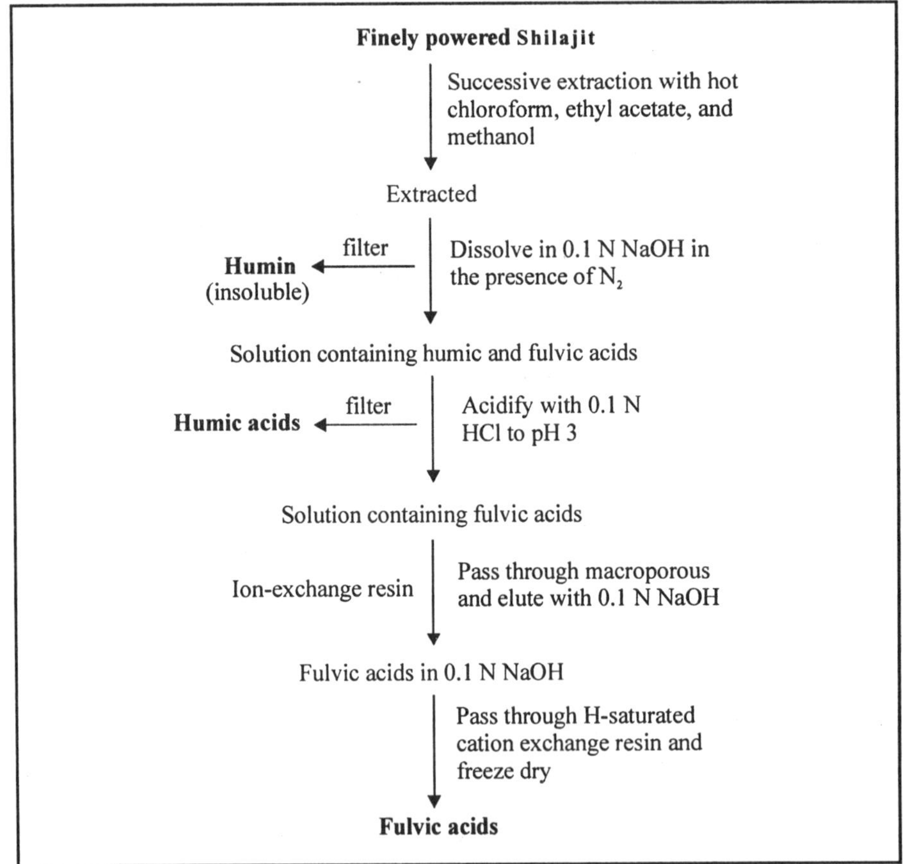

Fig. 5.2: Schematic representation of the modified method used by Khanna (2005) for separation of humic substances from shilajit

in 0.1 N aqueous sodium hydroxide with intermittent shaking in the presence of nitrogen at room temperature for 24 hours. The suspension was filtered to remove humin (insoluble at all pH in water) and the filtrate was acidified with dilute HCl to a pH ~3. The solution was allowed to stand at room temperature (25°C) overnight. Humic acid, which separated out as coagulate, was filtered, dried, and pulverized. The filtrate thus obtained was shaken with macroporous ion-exchange resin in order to adsorb the FA which was then eluted using 0.1 N aqueous sodium hydroxide solution. The FA thus obtained in alkali was passed through hydrogen-saturated cation exchange resin in order to exchange the sodium ions with hydrogen ions. The final FA solution was concentrated and freeze dried to obtain amorphous FA.

Table 5.1 compares the yield of fulvic acids from different shilajit samples obtained by the earlier reported method (Ghosal

Table 5.1: Comparison of the yield of fulvic acid from shilajit of different sources

Shilajit source	**% Yield of fulvic acid**							
	Method reported by Ghosal *et al.* (1989)				**Method developed by Khanna (2005)**			
	I	**II**	**III**	**Mean±SD**	**I**	**II**	**III**	**Mean±SD**
Rock shilajit (RS)	0.4	0.7	1.0	0.7 ± 0.3	5.8	5.0	4.9	5.2 ± 0.5
Gurukul Kangri (GK)	1.1	1.8	1.5	1.5 ± 0.4	11.1	12.5	10.5	1.4 ± 1.0
NR Enterprises (NR)	2.5	2.7	2.8	2.7 ± 0.2	12.2	13.5	13.0	2.9 ± 0.7
Pioneer Enterprises (PE)	2.4	3.1	3.6	3.0 ± 0.6	13.9	12.5	13.3	3.2 ± 0.7

et al., 1989) and the improved method developed by Khanna (2005). As can be seen, there was a significant improvement in the yield of fulvic acids using the new method. In addition, the fulvic acids obtained were completely soluble in water as against the earlier reported method.

The humic and fulvic acids extracted from various shilajit samples were characterized for their physical, chemical, and spectral characteristics (Khanna, 2005).

CHARACTERIZATION OF HUMIC ACIDS

Humic acids extracted from various shilajit samples were characterized based on their physicochemical properties such as color, odor, taste, solubility in water, and other organic solvents, etc. Scanning electron microscopy, elemental and spectral analysis such as UV, FTIR, and XRD diffraction were performed. Thermal properties of the samples were analyzed by means of DSC thermograms. The spectral properties were also compared with that of a standard soil humic acid sample from Sigma Aldrich.

Table 5.2 lists the physical characteristics of humic acids extracted from shilajit of different origin sources. All the HA samples were brownish black in color and had a typical characteristic odor and taste (Figure 5.3).

The pH of a 2% aqueous solution ranged from about 3.46 to 3.86. The E_4/E_6 ratio for all the HA samples ranged from about

Fig. 5.3: Humic acid extracted from shilajit

Table 5.2: Comparison of the physical characteristics of humic acid from shilajit of different origin sources

Characteristic	Humic acid (RS)	Humic acid (GK)	Humic acid (NR)	Humic acid (PE)
Nature	Dark brown powder	Dark brown powder	Dark brown powder	Dark brown powder
Color	Dark brown	Dark brown	Dark brown	Dark brown
Odor	Characteristic	Characteristic	Characteristic	Characteristic
Taste	Characteristic	Characteristic	Characteristic	Characteristic
pH of 2% aq Sol.	3.86	3.77	3.46	3.68
Absorbance at 465 nm (E_4)	0.513	0.542	0.284	0.222
Absorbance at 665 nm (E_6)	0.144	0.180	0.072	0.072
E_4/E_6 ratio	3.56	3.01	3.94	3.08

3.0 to 4.0 which are consistent with those reported in the literature (Chen *et al.*, 1977).

ELEMENTAL ANALYSIS

Elemental analysis of humic substances is generally used to estabilish their nature and origin (Mcdonnel *et al.*, 2001). The C, H, N, and S content (% dry weight) for various humic acid samples extracted from different shilajit sources was analyzed by packing the humic acid powder in tin boats after careful weighing with the help of elemental analyzer (CHNS analyzer, Vario EL-III) and the results were compared with that of a standard humic acid available from Sigma Aldrich.

Table 5.3 compares the carbon, hydrogen, nitrogen, and sulfur contents of humic acids extracted from different sources of shilajit with those of soil humic acid standard from Sigma Aldrich. Each elemental composition was expressed as $\%g^{-1}$ dry weight powder.

The carbon content in the humic samples from different sources ranged from 27.4% to 51.4%; hydrogen content, from 2.9% to

Table 5.3: Elemental analysis of humic acids extracted from shilajit

Sources of humic acids	%C	%H	%N	%S	C/N ratio
Rock shilajit (Dabur)	36.46	5.15	3.03	0.70	12.00
Shudh shilajit (Gurukul Kangri)	45.36	5.92	2.31	0.39	19.63
Shilajit extract (Natural Remedies)	51.48	5.89	3.27	0.81	15.73
Shilajit extract (Pioneer Enterprises)	27.44	2.90	1.24	0.26	22.10
Humic acid standard (Sigma Aldrich)	42.28	4.25	0.57	0.81	73.09

5.9%; nitrogen content from 1.24% to 3.27%, and sulfur content, from 0.26% to 0.87%. Interestingly, while the variation in the contents was observed in various samples, the C/N ratio for the samples was in a narrow range of 12 to 22. In contrast, the C/N ratio in the humic acid standard was found to be 73 mainly resulting from a low nitrogen content.

UV–Vis Spectroscopy

UV-vis spectra of various HA extracted from different origin shilajit were obtained on a Shimadzu, 1601 UV–VIS spectrophotometer by dissolving the various HA samples in water and recording the spectra in a 1-cm quartz cuvette by scanning from 200 to 800 nm and the obtained results are shown in Figure 5.4. Since humic substances usually yield uncharacteristic spectra in the UV and visible, E_4/E_6 ratio (ratio of the absorbance of the solution at 465 and 665 nm) (Schnitzer, 1972) was determined for the various samples. This ratio has been widely used by soil scientists for characterization purposes.

As can be seen from the figure, the samples did not exhibit any sharp maxima but exhibited a slight hump near 260–280 nm characteristic of humic substances (Schnitzer, 1972). As

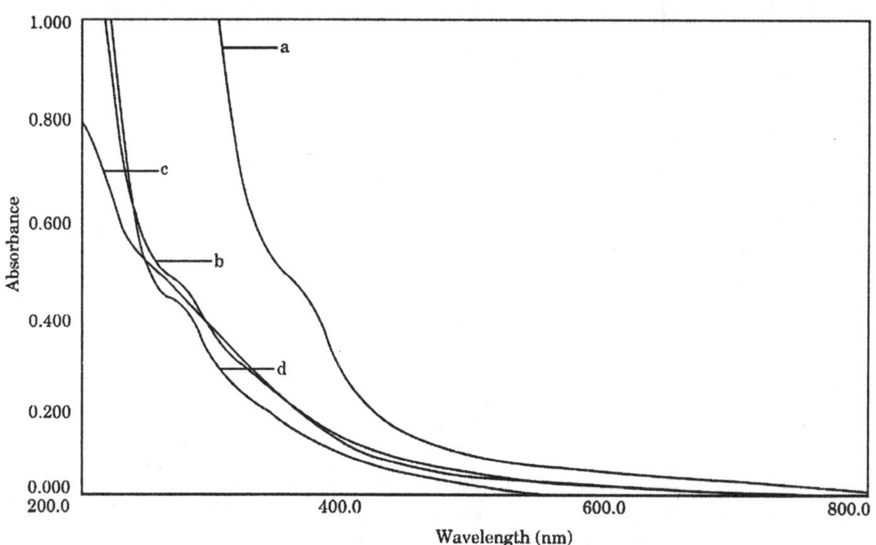

Fig. 5.4: UV–visible spectra of humic acid extracted from shilajit of different origins: (a) RS, (b) GK, (c) NR, and (d) PE

discussed earlier, this hump is attributed to the absorption of radiation by the double bonds C–C, C–C, and N–N of aromatic or unsaturated components of humic substances (Domeizel *et al.*, 2004). The variation in the hump observed with the different samples of shilajit could be attributed to the variation in the concentration of aromatic compounds which in turn is characteristic of the difference in the humification process of the different shilajit samples.

Fourier Transform Infra-red SPECTROSCOPY (FTIR)

The infrared spectra of various humic acid samples were recorded on a FTS 40 (BioRad, USA) FTIR instrument and WIN IR software by the KBr pellet technique. Two milligrams of previously dried sample was mixed with 100 mg KBr and compressed into a pellet on an IR hydraulic press. These pellets were made immediately prior to the recording of the spectrum. Scanning was done from 4000 to 450 cm^{-1} and the results obtained are shown in Figure 5.5.

As can be seen from the figure, FTIR spectra of humic acids

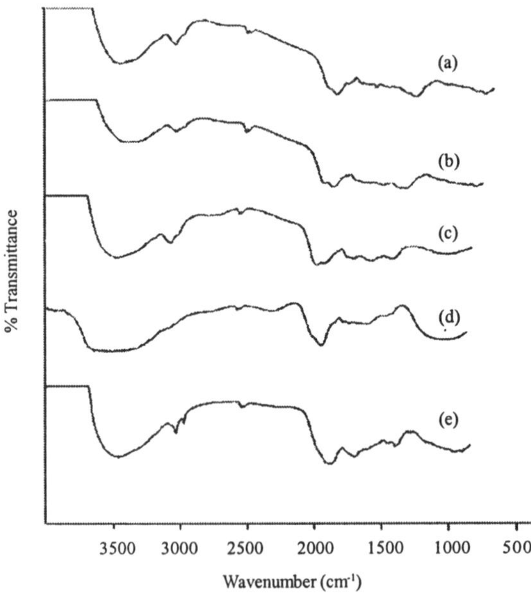

Fig. 5.5: FTIR spectra of humic acid extracted from shilajit of different origins: (a) RS, (b) GK, (c) NR, (d) PE, and (e) Laurentian humic acid

were characterised by a relatively few bands that were broad. All humic acid samples exhibited broad bands at about 3400 cm^{-1}, 1725 cm^{-1}, and 1630 cm^{-1} which can be attributed to hydrogen bonded OH group, C-O stretching of COOH, and C-C double bond, respectively. Sharp bands were observed in the region of 2925 cm^{-1}, 1400 cm^{-1}, and 1050 cm^{-1} which can be attributed to the bending vibration of aliphatic C-H group, the O-H bending vibrations of alcohols or carboxylic acids and OH bending deformation of carboxyl groups, respectively (Schnitzer, 1972).

Powder X-ray Diffraction

Powder X-ray diffraction patterns of powdered samples of HA were obtained using a Panalytical X-ray diffractometer PW3719. All the samples were treated according to the following specifications:

Target/filter (monochromator)	:	Cu
Voltage/current	:	40 kV/50 mA
Scan speed	:	4°/min
Smoothing	:	0

A typical X-ray diffractogram of humic acid extracted from rock shilajit is shown in Figure 5.6. The sample largely exhibited a noncrystalline nature as evident from the absence of sharp diffraction peaks. A few peaks were, however, observed which could be attributed to the presence of traces of lipid crystallites, clay particles, and other crystalline and microcrystalline materials.

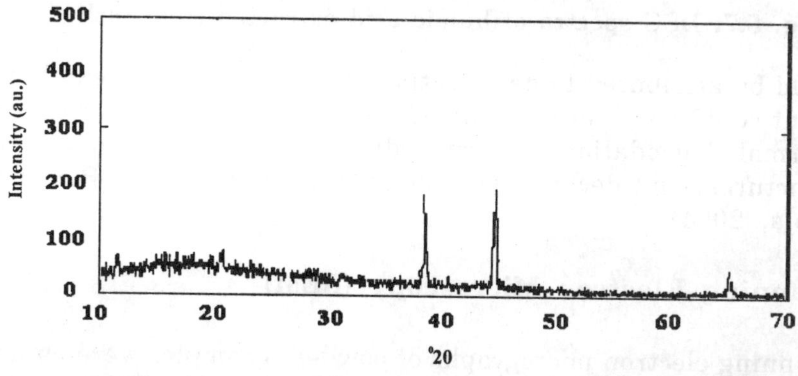

Fig. 5.6: XRD spectra of humic acid extracted from rock shilajit

The behavior is consistent with the behavior observed in case of humic substances from other sources (Chilom and Rice, 2005; Visser and Mendel, 1971).

Differential Scanning Calorimetry (DSC)

A Perkin-Elmer Pyris 6 instrument was used for recording DSC thermogram of the HA samples obtained from different shilajit sources. Samples (2–8 mg) were accurately weighed and heated in closed aluminium crimp cells at a rate of 10 °C/min under nitrogen purge with a flow rate 20 mL/min over the 50-300 °C temperature range.

Humic acid samples did not exhibit any sharp endotherm indicating that it does not have any defined melting point (Figure 5.7). A shallow endotherm could be observed near 100 °C which

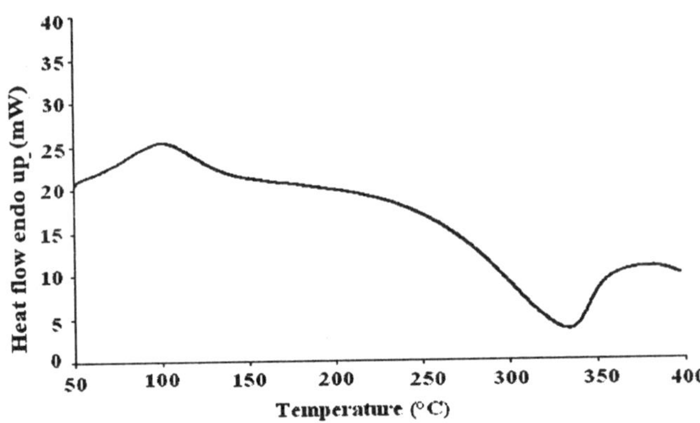

Fig. 5.7: DSC spectra of humic acid extracted from rock shilajit

could be attributed to dehydration of the samples. An exothermic event could be observed near 330 °C which could be attributed to thermal degradation of carbohydrates, dehydration of aliphatic structures, and decarboxylation of carboxylic groups (Pietro and Paola, 2004).

Scanning Electron Microscopy (SEM)

Scanning electron micrographs of powdered samples were obtained using a Joel JSM-840 Scanning Microscope with a 10 kV accelerating voltage. The surface of samples for SEM was made

SEM – Humic acid from Rock Shilajit-500x

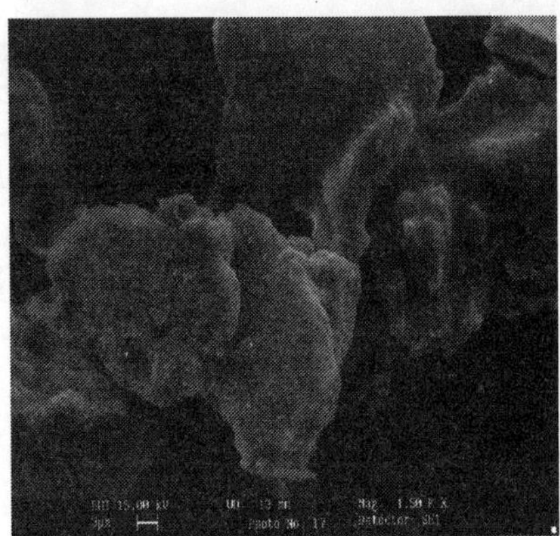

SEM – Humic acid from Rock Shilajit-1500x

Fig. 5.8: Scanning electron micrographs of humic acid from rock shilajit

electrically conductive in a sputtering apparatus (Fine Coat Ion Sputter JFC-1100) by evaporation of gold. Scanning electron micrographs of humic acids extracted from rock shilajit taken at magnifications of 1500× and 3000× are shown in Figure 5.8. Humic acids mainly occurred as loose spongy structures with particles tending to aggregate to each other.

CHARACTERIZATION OF FULVIC ACIDS

Fulvic acids from various shilajit samples were extracted and characterized based on their physicochemical properties such as color, odor, taste, solubility in water, and other organic solvents, etc. Scanning electron microscopy, elemental and spectral analysis such as UV, FTIR, ^{1}H NMR, and XRD diffraction as well as elemental composition analysis by FT-ICR mass spectrometry were performed. Thermal properties of the samples were analyzed by means of DSC thermograms. The spectral properties were also compared with that of a standard soil fulvic acid sample (Laurentian fulvic acid, Fredriks Research Products, the Netherlands).

Table 5.4 lists the physical characteristics of fulvic acids extracted from shilajit of different origin sources. All the samples were yellowish brown in color (Figure 5.9) and had a characteristic odor and taste.

Table 5.4: Comparison of the physical characteristics of fulvic acid from shilajit of different sources

Characteristic	FA (RS)	FA (GK)	FA (NR)	FA (PE)
Nature	Yellowish brown powder	Yellowish brown powder	Yellowish brown powder	Yellowish brown powder
Color	Yellowish brown	Yellowish brown	Yellowish brown	Yellowish brown
Odor	Characteristic	Characteristic	Characteristic	Characteristic
Taste	Characteristic	Characteristic	Characteristic	Characteristic
pH of 2% aq Sol.	3.04	3.44	3.09	3.46
Absorbance at 465 nm (E_4)	0.192	0.239	0.358	0.319
Absorbance at 665 nm (E_6)	0.066	0.103	0.152	0.116
E_4/E_6 ratio	2.91	2.32	2.36	2.75

Fig. 5.9: Fulvic acid extracted from shilajit

The pH of a 2% aqueous solution ranged from about 3.0 to 3.5, while the E_4/E_6 ratio for all the samples ranged from about 2.3 to 3.0 which are consistent with those reported in the literature (Chen *et al.,* 1977).

UV-Vis Spectroscopy

UV-vis spectra of fulvic acids extracted from different origin–shilajit were obtained on a Shimadzu, 1601 UV–VIS spectrophotometer by dissolving the various fulvic acids samples in water and recording the spectra in a 1 cm quartz cuvette by

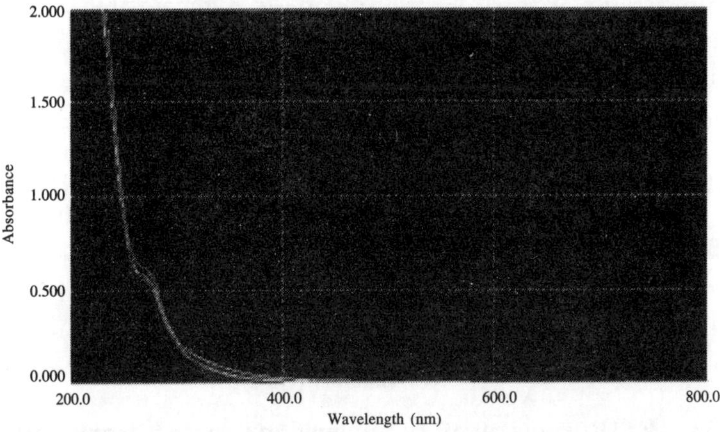

Fig. 5.10: UV-visible spectra of fulvic acid extracted from shilajit of different origins: (a) RS, (b) GK, (c) NR, and (d) PE

scanning from 200 to 800 nm and the results are shown in Figure 5.10. Since humic substances usually yield uncharacteristic spectra in the UV and visible, E_4/E_6 ratio (ratio of the absorbance of the solution at 465 and 665 nm) (Schnitzer, 1972) was determined for the various samples.

As shown in the figure, the samples did not exhibit any sharp maxima, however, a slight hump near 260–280 nm characteristic of humic substances was detected (Schnitzer, 1972). The variation in the hump observed with the different samples of fulvic acids could be attributed to the difference in the maturity and hence humification of different samples of shilajit which these were extracted from.

Fourier Transform Infra-red Spectroscopy

The infrared spectra of Shilajit samples were recorded on a FTS 40 (BioRad, USA) FTIR instrument and WIN IR software by the KBr pellet technique. Two mg of previously dried sample was mixed with 100 mg KBr and compressed into a pellet on an IR hydraulic press. These pellets were made immediately prior to the recording of the spectrum. Scanning was done from 4000 to 450 cm^{-1} and the results are shown in Figure 5.11.

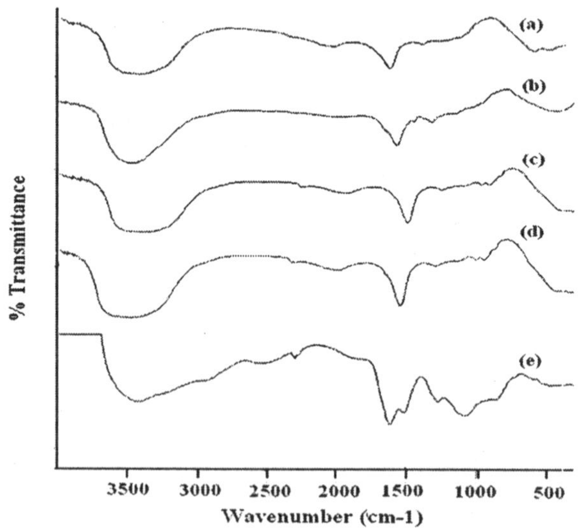

Fig. 5.11: FTIR spectra of fulvic acid extracted from shilajit of different origins: (a) GK (b) PE (c) RS (d) NR (e) Laurentian fulvic acid

As seen in the figure, the FTIR spectra of the different samples of fulvic acid extracted from shilajit of varying origin as well as that of the laurentian fulvic acid standard are shown. All the samples of fulvic acids extracted from shilajit of varying origin exhibited a distinct similarity and also shared a number of common features with the laurentian fulvic acid standard extracted from laurentian soil. All the samples exhibited a broad band at about 3400 cm^{-1} which can be attributed to hydrogen bonded OH group. Sharp bands were observed in the region of 1640 cm^{-1} (conjugated C-C double bond), 1400 cm^{-1} (O-H bending vibrations of alcohols or carboxylic acids) and 1140 cm^{-1} (C-O stretching of polysaccharide or polysaccharide like substances) (Schnitzer, 1972).

Fourier Transform Ion Cyclotron Resonance Mass Spectrometry (FT-ICR MS) (Khanna *et al.*, 2008)

An ultrahigh resolution mass spectra was obtained for fulvic acid sample extracted from rock shilajit using an Apex Qe 9.4T FT-ICR mass spectrometer (Bruker Daltonic Inc., Billerica, MA, USA). The fulvic acid sample was desalted before mass spectrometric analysis using a styrene divinyl benzene polymer type adsorber (Varian, PPL) and the extraction solvent methanol. After extraction of the fulvic acids from the adsorber the final concentration was about 1 mg/mL. The methanol extract was diluted with same volume of Milli-Q water for electrospray measurements. The mass spectrum of the fulvic acids sample was performed in negative ion mode with a detection range between *m/z* 210 and *m/z* 1000. 2M datapoints were used for data acquisition resulting in a resolution of about 300.000 at *m/z* 400. The spectrum was calibrated internally with fatty acids. The elemental composition analysis was performed with a script of the data processing software for automatic assignment of all mass peaks in the spectrum using a mass tolerance of 1 ppm.

The Kendrick plot analysis of the mass spectrometric data is shown in Figure 5.12. The Kendrick mass defects increased consistently with the Kendrick mass. This is a typical behavior of humic and fulvic acids due to the increase of oxygen atoms with the Kendrick mass. The linear regression of the data gave an intercept of 0.02 and a slope of 7.7×10^{-4}. The center of mass values of this measurement were *m/z* 395 of the Kendrick mass

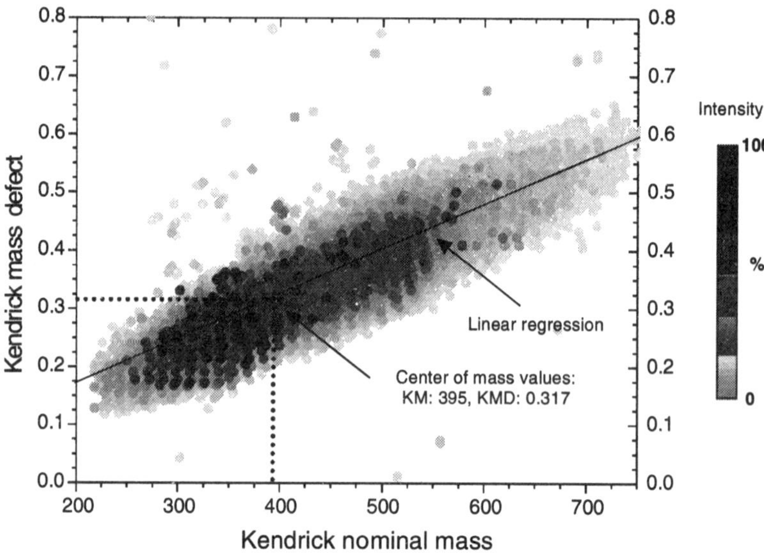

Fig. 5.12: Kendrick plot of the FT-ICR MS electrospray spectrum in negative ion mode of fulvic acids extracted from rock shilajit

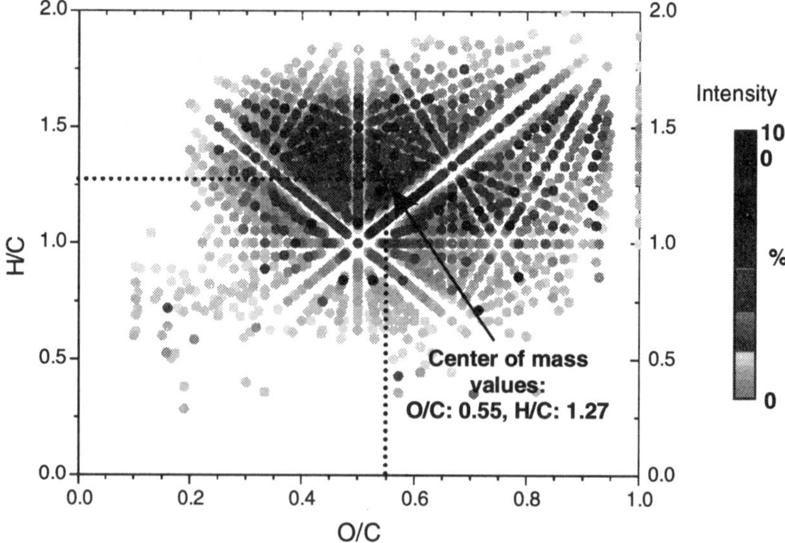

Fig. 5.13: Van Krevelen plot of identified nitrogen free compounds of the FT-ICR MS electrospray spectrum in negative ion mode of fulvic acids extracted from rock shilajit

and 0.317 of the Kendrick mass defect. The elemental composition analysis of nitrogen free compounds is shown in Figure 5.13 using the van Krevelen plot approach. The center of mass value of the O/C ratio of 0.55 indicates a relatively high rate of oxygen of the fulvic acids in shilajit. However, the high center of mass value of the H/C ratio of 1.27 demonstrates a relatively low aromaticity of these compounds. The average molecular formula of the nitrogen free elemental compositions measured by FT-ICR mass spectrometry in electrospray negative ion mode is $C_{18.2}H_{23.0}O_{10.0}$. The analysis of elemental compositions with one or two nitrogen atoms gave similar results with slightly different center of mass values of O/C and H/C ratios.

Nuclear Magnetic Resonance Spectrometry

Proton Magnetic Resonance (^1H NMR) spectra for the purified fulvic acid sample from rock shilajit were recorded on Brucker Model DRX-300 NMR spectrometer in $DMSOd_6$ and D_2O using tetramethylsilane (TMS) as the internal standard and the results are shown in Figures 5.14 and 5.15, respectively.

As can be seen from the Figure 5.12, ^1H NMR spectra of fulvic acid in $DMSOd_6$ exhibited various signals at chemical shifts, 0.89–2.17 ppm, suggesting the presence of aliphatic multiplet protons (methyl protons, methylene protons, and protons present on aliphatic carbons which are two or more carbons away from

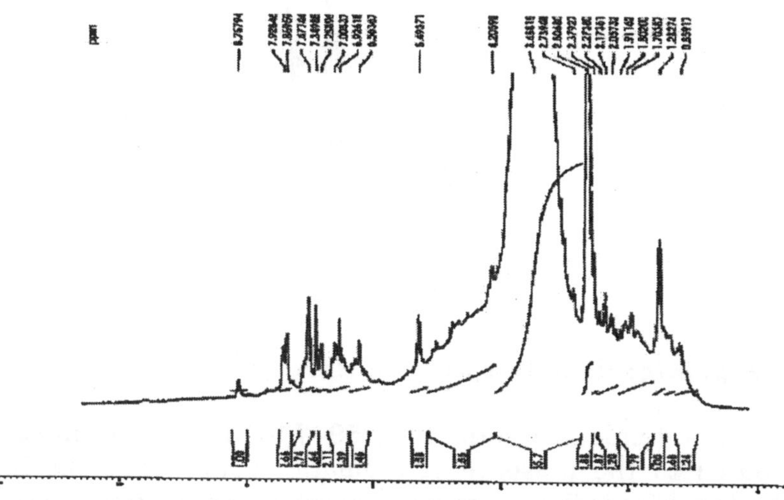

Fig. 5.14: ^1H NMR spectra of shilajit fulvic acid in $DMSOd_6$

aromatic rings or polar functional groups). Signal due to $DMSOd_6$ solvent was found at 2.50 ppm. A broad absorption signal at 3.45 ppm was found due to presence of moisture in $DMSOd_6$. A sharp signal at 5.49 ppm was found due to OH group in sugar, which strongly supports the presence of carbohydrate molecules in the shilajit fulvic acid. Signals in 6.56–7.92 ppm region is due to presence of multiplet aromatic protons. There is broad singlet at 8.75 ppm indicating the presence of N-containing molecules in shilajit fulvic acid. Protons in the aliphatic region is five times more as compared to aromatic region, suggesting aliphatic rich structure of shilajit fulvic acid.

As can be seen from Figure 5.15, ^{1}H NMR spectra of fulvic acid in D_2O also exhibited various signals at chemical shifts, 0.89–4.39 ppm, suggesting the presence of aliphatic protons (methyl protons, methylene protons, and protons present on aliphatic carbons which are two or more carbons away from aromatic rings or polar functional groups). Signals of carbohydrate and N-containing protons were absent due to exchange of protons with

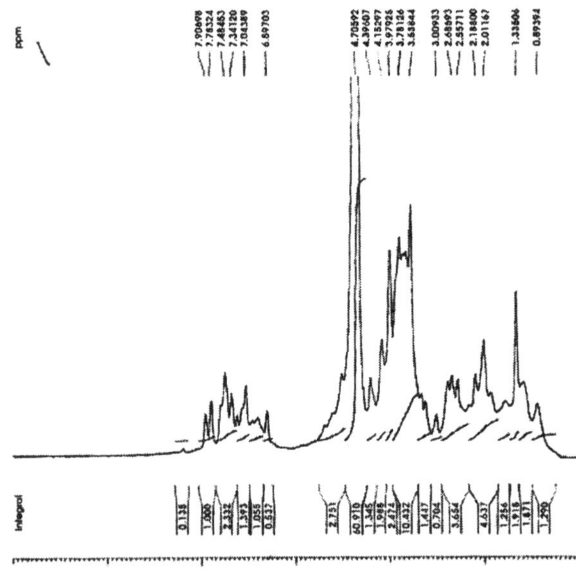

Fig. 5.15: ^{1}H NMR spectra of shilajit fulvic acid in D_2O

deuterium oxide. Signals in 6.59–7.90 ppm region can be attributed to the presence of aromatic protons. Protons in the aliphatic region is five times more as compared to aromatic region, suggesting aliphatic rich structure of shilajit fulvic acid.

Powder X-ray Diffraction

Powder X-ray diffraction patterns of powdered samples of FA were obtained using a Panalytical X-ray diffractometer PW3719. All the samples were treated according to the following specifications:

Target/filter (monochromator)	:	Cu
Voltage/current	:	40 kV/50 mA
Scan speed	:	4°/min.
Smoothing	:	0

A typical X-ray diffractogram of FA extracted from rock shilajit is shown in Figure 5.16. The sample largely exhibited a

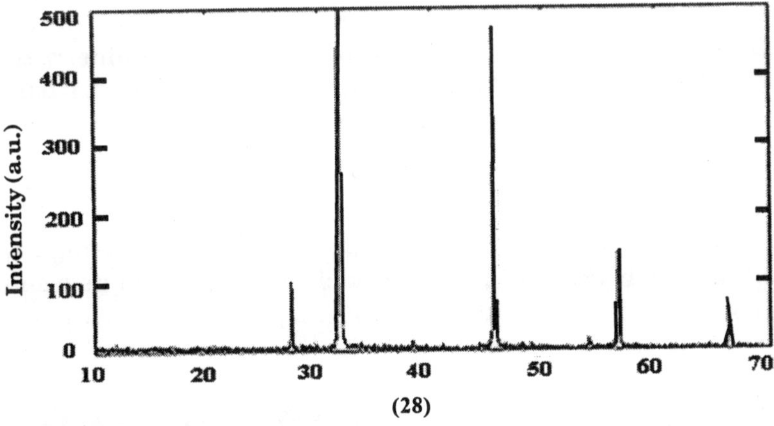

Fig. 5.16: XRD spectra of fulvic acid from rock shilajit

noncrystalline nature as evident from the absence of sharp diffraction peaks. A few peaks were, however, observed which could be attributed to the presence of traces of lipid crystallites, clay particles, and other crystalline and microcrystalline materials. The behavior is consistent with the behaviour observed in case of humic substances from other sources (Chilom and Rice, 2005; Visser and Mendel, 1971).

Differential Scanning Calorimetry (DSC)

A Perkin-Elmer Pyris 6 instrument was used for recording DSC thermogram of the FA samples obtained from shilajit. Samples

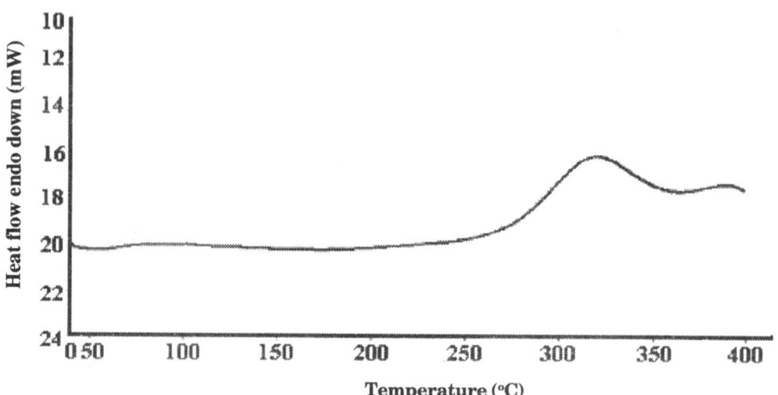

Fig. 5.17: DSC spectra of fulvic acid from rock shilajit

(2–3 mg) were accurately weighed and heated in closed aluminium crimp cells at a rate of 10 °C/min under nitrogen purge with a flow rate 20 mL/min over the 50–300 °C temperature range.

FA sample did not exhibit any sharp endotherm indicating that it does not have any defined melting point (Figure 5.17). An exothermic event could, however, be observed at a temperature above 250 °C which could be attributed to the thermal degradation of carbohydrates, dehydration of aliphatic structures, and decarboxylation of carboxylic groups (Pietra and Paola, 2004).

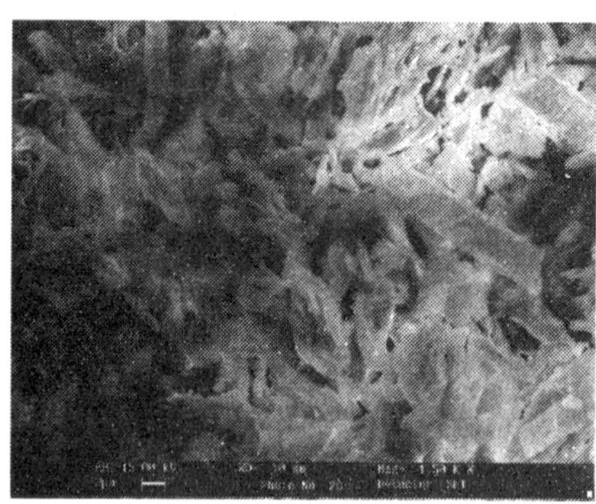

SEM: Fulvic acid from Rock Shilajit-1500x

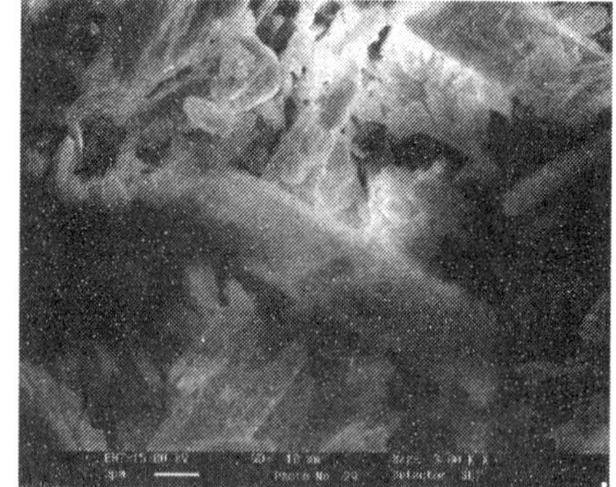

SEM – Fulvic acid from Rock Shilajit-3000x

Fig. 5.18: Scanning electron micrographs of fulvic acid from rock shilajit

Scanning Electron Microscopy (SEM)

Scanning electron micrographs of powdered samples were obtained using a Joel JSM-840 Scanning Microscope with a 10 kV accelerating voltage. The surface of samples for SEM was made electrically conductive in a sputtering apparatus (Fine Coat Ion Sputter JFC-1100) by evaporation of gold. Figure 5.18 shows Scanning electron micrographs of fulvic acids extracted from rock shilajit at resolution of 1500× and 3000×. Fulvic acids occurred mainly as elongated fibres woven into a loose mesh-like structure. The results are consistent with the reported observations for fulvic acids extracted from soil (Schnitzer, 1978).

6

SURFACTANT PROPERTIES OF HUMIC SUBSTANCES

Humic substances are not only predominantly hydrophilic (except at lower acidic pH) but they also contain a substantial concentration of aromatic rings, fatty acid esters, aliphatic hydrocarbon, and other hydrophobic substances, which together with the hydrophilic groups account for the surface activity of these materials. The hydrophilic oxygen containing functional groups (COOH, C=O, OH) are thought to play a significant role in lowering the surface tension of water and there by increasing aqueous wettability of hydrophobic materials. Tschapek and Wasowki (1976) were amongst the first to demonstrate the surfactant properties of humic substances.

It has also been recognized that the presence of even a small amount of humic acid in an aqueous solution can significantly enhance the water solubility of a hydrophobic organic compound (Gaffney *et al.*, 1996). This solubilization in solutions is often attributed to the presence of micelles. The structure of humic acids is such that it allows them to function as surfactants with the ability to bind both hydrophilic and hydrophobic materials. This function in combination with their colloidal properties makes humic acids effective agents in transporting both organic and inorganic materials in the environment.

Humic acids being highly aromatic as compared to fulvic acids become insoluble at low pH values when the carboxylate groups become protonated that may also lead to formation of intramolecular "pseudomicelles", as opposed to intermolecular micelles, because of coiling and contraction of humic acid chains (Wandruszka *et al.*, 1998). Pseudomicelles are submicroscopic aggregates of humic acid molecules that are analogous to the micelles formed by soaps and other surface-active compounds. As such, they have nonpolar cores, comparable to miniature oil drops, and polar surfaces that make them water compatible. Their structure in humic acid is less defined than it is in synthetic detergents, due to variations in molecular size and composition of humic acid. The effects, however, are similar. It is found that humic acid pseudomicelles can be formed by both intra- and intermolecular processes. In the intramolecular case, humic acid polymers coil and fold to create molecular domains that may be likened knots in a string.

The surfactant properties of humic and fulvic acids extracted from shilajit were investigated by determining the effect of increasing concentration of humic and fulvic acids on the surface tension of water. The surface tension of the solutions was determined by the drop-weight method using a stalagmometer. Solutions of humic or fulvic acids in the concentration range of 0% to 1.4% (w/v) were prepared. Each solution was separately sucked into the stalagmometer and allowed to drop slowly from it. The drop rate was adjusted to approximately 2–3 drops/min and the weight of 10 drops was determined. The determination was repeated twice. Surface tension of the solution was calculated using the formula:

$$\text{Surface tension of humic/fulvic acid solution} = \frac{\text{Surface tension of water}}{\text{Weight of water}} \times \text{Weight of fulvic acid solution}$$

SURFACTANT PROPERTIES OF HUMIC ACIDS

The effect of increasing concentration of extracted humic acids on the surface tension of water is shown in Table 6.1 and Figure 6.1.

It is shown in Figure 6.1 that increasing the concentration of extracted humic acids in water clearly led to a decrease in its surface tension. The decrease was gradual initially till a concentration of about 0.8% (w/v) was reached after which it

Table 6.1: Effect of increasing concentration of humic acids on the surface tension of water

Conc. of humic acid (%;w/v)	Mean weight of 10 drops (g)	Surface tension (dyn/cm)
0.0	1.632	72.80
0.1	1.607	71.67
0.2	1.596	71.18
0.3	1.562	69.67
0.4	1.539	68.64
0.5	1.522	67.88
0.6	1.466	65.38
0.7	1.365	60.88
0.8	1.290	57.53
0.9	1.297	57.85
1.0	1.305	58.20
1.1	1.297	57.85
1.2	1.296	57.80
1.3	1.307	58.29
1.4	1.290	57.53

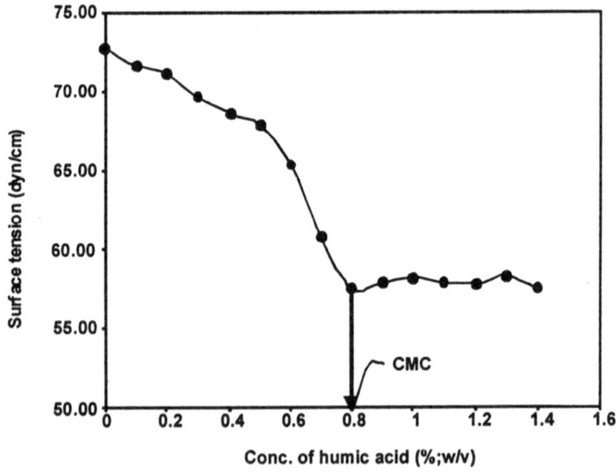

Fig. 6.1: Effect of humic acids on the surface tension of water

rose slightly and then became almost constant. This could be due to the formation of micelle at this concentration. This demonstrates that humic acids extracted from shilajit indeed possess surfactant properties. The value of 0.8% (w/v) for the critical micelle concentration (CMC) is in agreement with the reported value of 0.7% (w/v) for humic acids extracted from soil (Gaffney *et al.*, 1996).

SURFACTANT PROPERTIES OF FULVIC ACIDS

Table 6.2 and Figure 6.2 show the effect of increasing concentration of extracted fulvic acids on the surface tension of water.

Figure 6.2 shows that, increasing the concentration of extracted fulvic acids in water led to a sharp decrease in its surface

Table 6.2: Effect of increasing concentration of fulvic acids on the surface tension of water

Conc. of fulvic acid (%;w/v)	Mean weight of 10 drops (g)	Surface tension (dyn/cm)
0.0	1.830	72.80
0.1	1.484	59.03
0.2	1.233	49.05
0.3	1.190	47.34
0.4	1.088	43.28
0.5	1.045	41.57
0.6	1.014	40.34
0.7	1.020	40.58
0.8	0.989	39.34
0.9	0.916	36.44
1.0	0.913	36.32
1.1	0.906	35.80
1.2	0.892	35.48
1.3	0.883	35.13
1.4	0.877	34.89

tension. In fact, the decrease was comparatively more than that with humic acids. The decrease was gradual initially till a concentration of about 0.3% (w/v) was reached after which it rose slightly and then became almost constant or declined slowly. The relationship between the surface tension and concentration of fulvic acid could be divided into two linear sections as shown by the dotted lines in Figure 6.2 and the intersection of the two lines could be regarded as the critical micelle concentration (CMC). This CMC of 0.3% (w/v) is lower than the CMC reported for fulvic acids extracted from soil (0.68%; w/v) (Gaffney *et al.*, 1996). This demonstrates that fulvic acids extracted from shilajit indeed possess greater surfactant properties than those extracted from other sources.

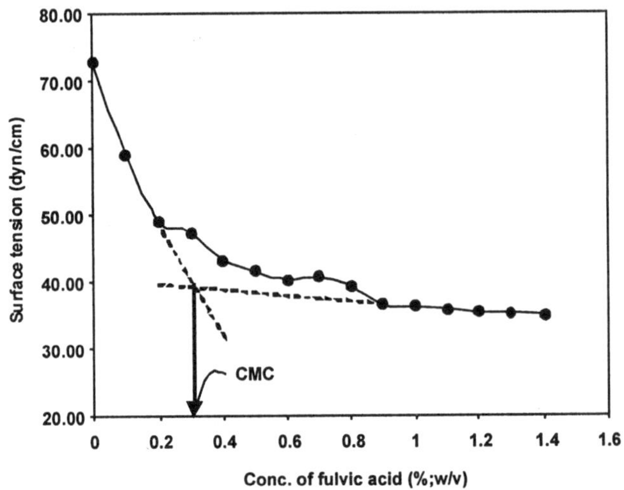

Fig.6.2: Effect of fulvic acids on the surface tension of water

MICELLAR SOLUBILIZATION OF DRUGS

I. Itraconazole

Itraconazole is a newer, broad spectrum, triazole antifungal drug, which is indicated for the treatment of dermatophytosis, candidiasis, mycosis, and a number of other superficial and systemic fungal infections in humans. Itraconazole has very poor aqueous solubility which often limits its bioavailability and poses a major challenge for formulating the drug into a suitable dosage form.

In order to investigate whether humic substances extracted from shilajit were capable of increasing the solubility of poorly soluble drugs due to micellar solubilization, Khanna (2005) studied the effect of increasing concentration of humic and fulvic acids on the aqueous solubility of itraconazole by carrying out phase-solubility studies according to the method of Higuchi & Connors (1965). Aqueous solutions (10 mL) of humic or fulvic acids were prepared in the concentration range of 0% to 1.6% (w/v) An excess amount of itraconazole (about 20 mg) was added to each sample contained in a stoppered glass tube. The tubes were shaken on a mechanical shaker, equipped with a thermostatically controlled water bath, for 7 days at 25 ± 2 °C. After 7 days, all the samples were centrifuged and the supernatant was filtered through a 0.22–µm membrane filter and analyzed by HPLC method. The concentration of drug in the solution was determined from the calibration curve.

Table 6.3 and Figure 6.3 show the effect of increasing concentration of humic acids on the aqueous solubility of itraconazole at 25 °C.

Table 6.3: Phase solubility studies of itraconazole with extracted humic acids

Sl. No.	Conc. of humic acid (%;w/v)	Conc. of itraconazole in solution (µg/mL)
1	0.0	0.066
2	0.2	0.145
3	0.4	0.166
4	0.6	0.164
5	0.8	0.240
6	1.0	0.409
7	1.2	0.441
8	1.4	0.460
9	1.6	0.472

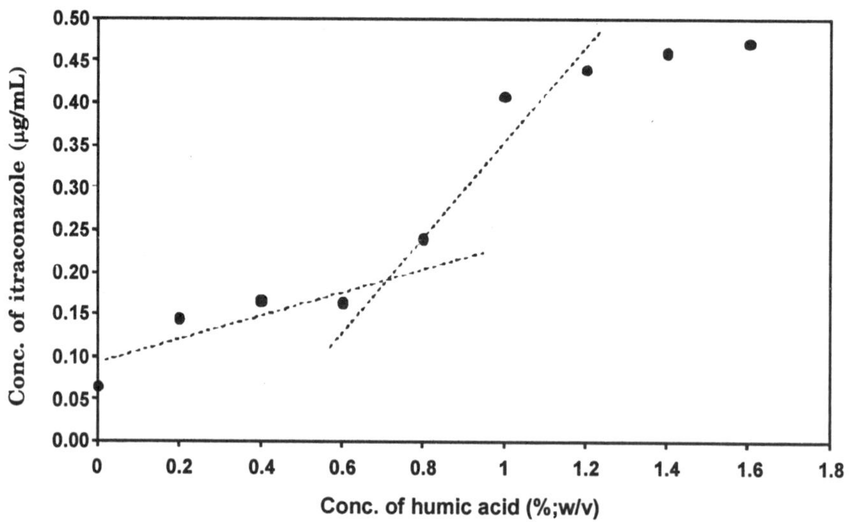

Fig. 6.3: Phase solubility studies of itraconazole with extracted humic acids

The phase solubility diagram in Figure 6.3 shows that extracted humic acids cause a slow increase in the solubility of itraconazole until a concentration of about 0.7% (w/v) where a sudden increase in the solubility of drug is noticed which could probably be due to localization of drug molecules in the hydrophobic domain of the micelles formed. This value of 0.7% (w/v) is in good agreement with the CMC of 0.8% (w/v) determined by the surface tension method and further supports the micelle-forming property of humic acids.

Table 6.4 and Figure 6.4 show the effect of increasing concentration of fulvic acids on the aqueous solubility of itraconazole at 25 °C.

Figure 6.4 shows that the extracted fulvic acids cause a slow increase in the solubility of itraconazole until a concentration of about 0.4% (w/v) is reached where a sudden increase in the solubility of drug is noticed. This could probably be due to localization of drug molecules in the hydrophobic domain of the micelles formed. This value of 0.4% (w/v) is in good agreement with the CMC of 0.3% (w/v) determined by the surface tension method and further supports the micelle forming property of fulvic acids.

Table 6.4: Phase solubility studies of itraconazole with extracted fulvic acids

Sl. No.	Conc. of fulvic acid (%;w/v)	Conc. of itraconazole in solution (µg/mL)
1	0.0	0.066
2	0.2	0.103
3	0.4	0.146
4	0.6	0.318
5	0.8	0.353
6	1.0	0.381
7	1.2	0.394
8	1.4	0.409
9	1.6	0.414

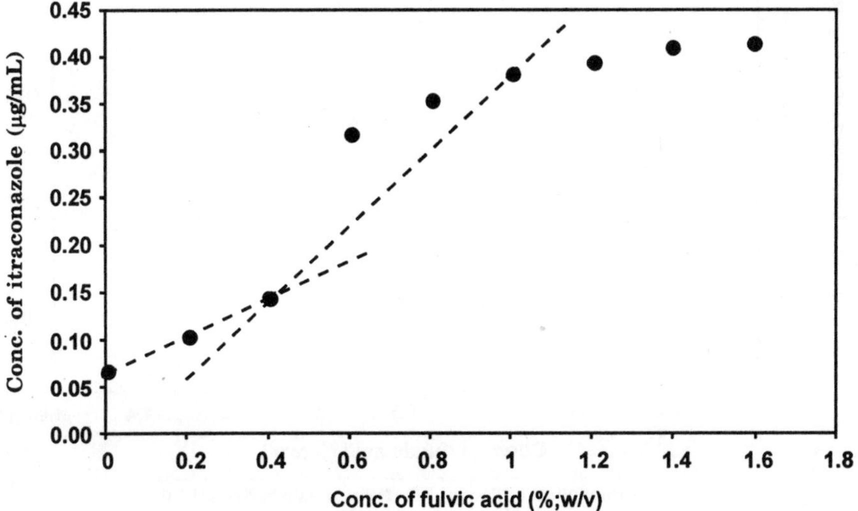

Fig. 6.4: Phase solubility studies of itraconazole with extracted fulvic acids

II. Ketoconazole

Ketoconazole is another imidazole antifungal agent used in the treatment of candidiasis, blastomycosis, and other systemic fungal infections in humans. Ketoconazole is a weak base with pK_a of 2.94 and 5.61. It is practically insoluble in water except at pH < 3. The major drawback associated with the therapeutic application of ketoconazole is its poor bioavailability upon oral administration because of its very low aqueous solubility and hydrophobic nature.

Karmarkar (2007) investigated the effect of increasing concentration of humic acid and fulvic acid on the aqueous solubility of ketoconazole by carrying out phase solubility studies at 25 °C in 0.1 M phosphate buffer of pH 5.0 and 6.0. The results of the study are shown in Table 6.5 and in Figure 6.5.

The phase solubility diagram (Figure 6.5) for ketoconazole shows that extracted humic acid causes a slow increase in the solubility of ketoconazole until a concentration of about 0.7% to 0.8% (w/v) where a sudden increase in the solubility of

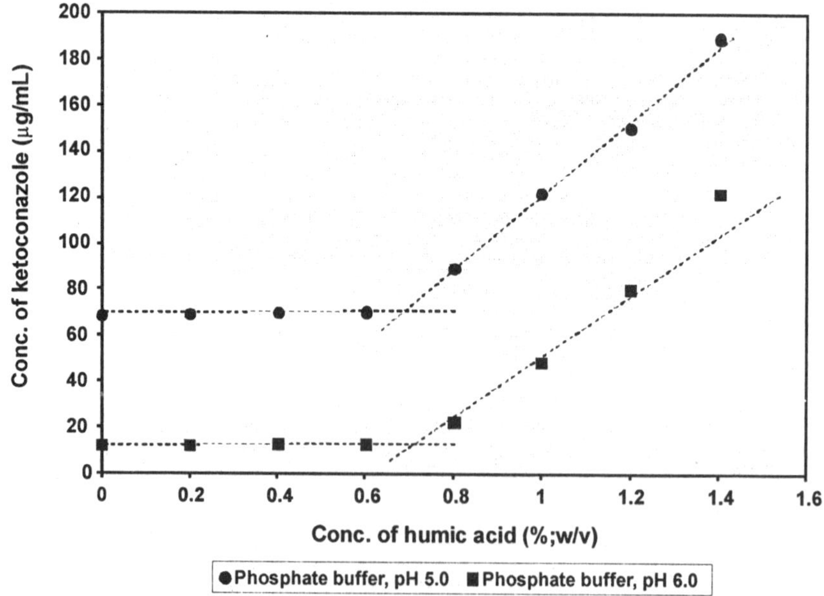

Fig. 6.5: Phase solubility study of ketoconazole with humic acid in 0.1 M phosphate buffer of pH 5.0 and 6.0

Table 6.5: Phase solubility studies of ketoconazole with extracted humic acid

Sl. No.	Conc. of humic acid (%; w/v)	Conc. of ketoconazole in phosphate buffer, pH 5.0 (μg/mL)	Conc. of ketoconazole in phosphate buffer, pH 6.0 (μg/mL)
1	0.0	69.0	12.0
2	0.2	69.8	12.5
3	0.4	70.2	12.8
4	0.6	70.6	12.8
5	0.8	89.7	22.9
6	1.0	122.4	48.9
7	1.2	150.7	80.5
8	1.4	189.7	122.6

Table 6.6: Phase solubility studies of ketoconazole with extracted fulvic acid

Sl. No.	Conc. of fulvic acid (%;w/v)	Conc. of ketoconazole in phosphate buffer, pH 5.0 (μg/mL)	Conc. of ketoconazole in phosphate buffer, pH 6.0 (μg/mL)
1	0.0	69.8	12.8
2	0.2	72.3	15.5
3	0.4	97.6	38.5
4	0.6	110.5	46.4
5	0.8	118.1	54.7
6	1.0	130.2	69.5
7	1.2	134.2	78.5
8	1.4	138.3	87.1

drug is observed which could probably be due to localization of drug molecule in the hydrophobic domain of the micelles formed. This value of 0.7% to 0.8% (w/v) is in good agreement with the CMC of 0.8% (w/v) determined by the surface tension method and further supports the micelle formation property of humic acid.

The effect of increasing concentration of fulvic acid on the aqueous solubility of ketoconazole was demonstrated by carrying out phase solubility studies at 25 °C in 0.1 M phosphate buffer of pH 5 and pH 6. The results are shown in Table 6.6 and in Figure 6.6.

Figure 6.6 shows that the extracted fulvic acid causes a slow increase in the solubility of ketoconazole until a concentration of about 0.3% to 0.4% (w/v) is reached where a sudden increase in the solubility of drug is observed. This could probably be due to localization of drug molecule in the hydrophobic domain of the micelle formed. The value of 0.3–0.4% (w/v) is in good agreement with the CMC of 0.3% (w/v) determined by the surface tension

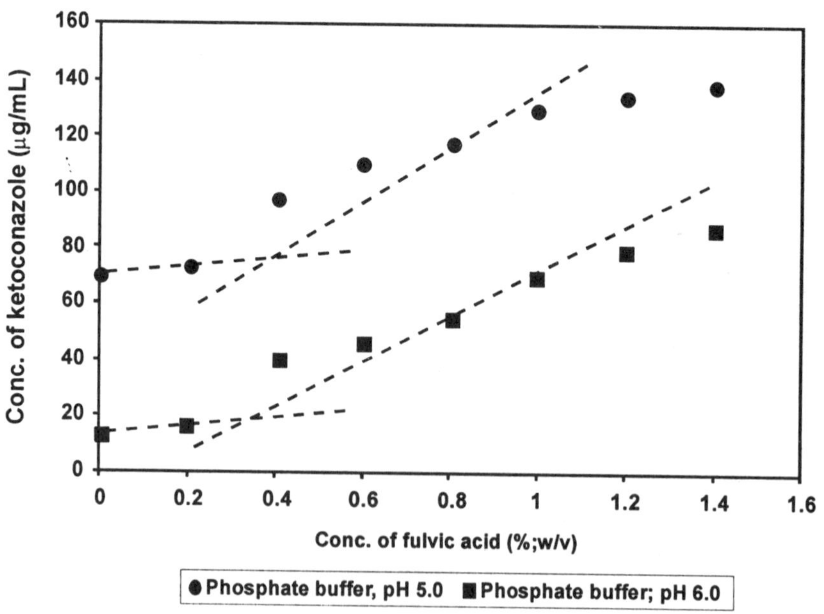

Fig. 6.6: Phase solubility study of ketoconazole with fulvic acid in 0.1 M phosphate buffer of pH 5.0 and 6.0

method and further supports the micelle forming property of fulvic acid.

III. Meloxicam

Meloxicam is a nonsteroidal anti-inflammatory drug prescribed for the long-term treatment of musculoskeletal complaints. It is a preferential COX-2 inhibitor with strong anti-inflammatory activity. It is practically insoluble in water. The poor aqueous solubility and wettability present problems for the preparation of pharmaceutical formulations with good release and nonvariable bioavailability. Long-term use of meloxicam has been implicated in causing gastric ulceration by interfering with the biosynthesis of prostaglandins and other arachadonic acid metabolite and also by remaining in contact with stomach mucosa for a longer duration of time.

Vashisht (2007) studied the effect of increasing concentration of fulvic acid on the aqueous solubility of meloxicam by carrying out the phase solubility study in water at 25 ± 2 °C. Concentration

Table 6.7: Phase solubility studies of meloxicam with extracted fulvic acids

Sl. No.	Conc. of fulvic acid (%;w/v)	Con. of meloxicam in solution (µg/mL)
1	0.0	10.0
2	0.1	13.0
3	0.2	16.2
4	0.3	19.5
5	0.4	36.6
6	0.5	38.7
7	0.6	40.9
8	0.7	43.0
9	0.8	44.9

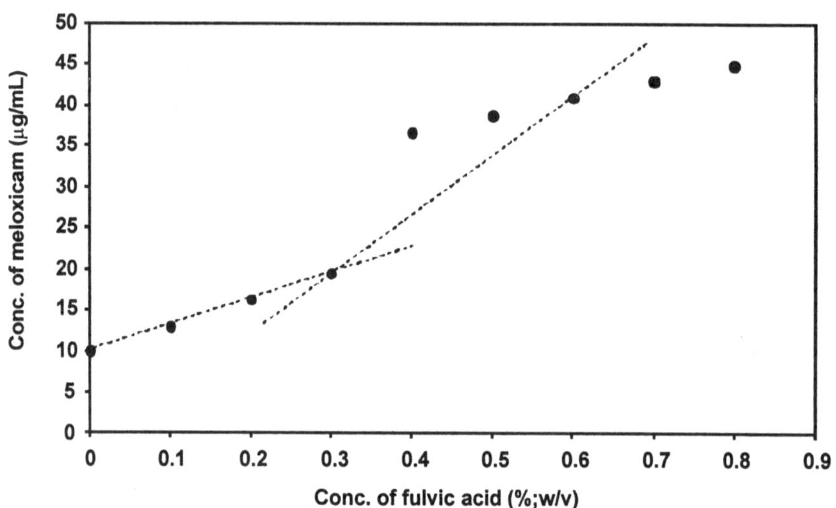

Fig. 6.7: Phase solubility studies of meloxicam with extracted fulvic acids

of fulvic acid was varied from 0.0% to 0.8% (w/v). The results are shown in Table 6.7 and Figure 6.7.

Figure 6.7 shows that the extracted fulvic acids cause a slow increase in the solubility of meloxicam until a concentration of about 0.3% (w/v) is reached where a sudden increase in the solubility of drug is noticed. This value of 0.3% (w/v) is in good agreement with the CMC of 0.3% (w/v) determined by the surface tension method and further supports the micelle-forming property of fulvic acids.

The above studies confirm the surfactant and micellar solubilization property of humic substances extracted from shilajit which can advantageously be used for increasing the solubility of a number of poorly soluble drugs.

7

DRUG COMPLEXING PROPERTIES OF HUMIC SUBSTANCES

One of the important characteristics of humic and fulvic acids is their pronounced chemical reactivity towards other molecules. Humic substances have an abundance of carboxyl groups and also possess weakly acidic phenolic groups. These groups contribute to the ion-exchange properties of humic materials. Humic substances are also known to be redox-active and possess free-radicals which can bind small molecules through both hydrogen bonding and nonpolar interactions. Humic substances exhibit both hydrophobic and hydrophilic characteristics and can bind to mineral surfaces. The abundance of oxygen-, nitrogen-, and sulphur-containing functional groups in these humic substances make them efficacious metal-complexing ligands (Livens, 1991). Given the variety of functional groups in humic molecules and the various ways in which they can interact with metals, a near infinite number of metal-humic complexes are possible in principle. It has long been recognized that natural organic matter is involved in the geochemical transport and concentration process of metal ions in the environment.

Extensive studies have shown that not much of the humic substances in soil are in free state but are mainly bound to colloidal clay (Nayak *et al.*, 1990). Humic acids combine in many ways with different fractions of naturally occurring matter and minerals:

(i) As salts of low-molecular weight organic acids (acetate, oxalate, lactate, and others).

(ii) As salts of humic substances with alkaline cations Humates.

(iii) As chelates with metal ions.

(iv) As substances held on clay mineral surfaces.

(v) The interaction of organic substances with clay has a multitude of consequences that are reflected in the physical, chemical, and biological properties of the soil matrix.

Several mechanisms are involved in the interaction of humic substances by clay minerals, the main ones being:

(i) van der Waals' forces

(ii) Bonding by cation bridging

(iii) Hydrogen bonding

(iv) Adsorption by association with hydrous oxides

(v) Adsorption on interlamellar spaces of clay minerals

Humic substances have also been implicated as the fundamental factor controlling the fate and transport of hydrophobic organic contaminants in soil and the subsurface environment (Bhandari *et al.*, 1996). The solubilization in water by humic substances of organic compounds, which are otherwise water insoluble is a matter of considerable interest. Wershaw *et al.* (1969) have shown that solubility of DDT in aqueous sodium humate solution is at least 20 times greater than that in water.

Chien *et al.* (1997) examined the "membrane micelle" model of humic substances in which micelle-like aggregates with hydrophobic interiors exist, in which nonpolar organic compound partition. Atrazine, labeled with trifluoromethane group on the ethylamino side chain, was solubilized in aqueous solutions of humic acid and F-19 NMR relaxation of atrazine induced by paramagnetic probes to humic acid solution was observed. The results confirmed that atrazine solubilized by humic acid occupies domain accessible only to neutral hydrophobic molecules and confirmed the existence of hydrophobic domains. It was suggested that atrazine resides in the interior of the humic acid micelles.

Studies carried out with humic and fulvic acids present in shilajit suggest that these have an "open" flexible structure perforated by voids of varying dimensions (Ghosal *et al.*, 1991). These voids are capable of entrapping bioactive molecules like the low-molecular-weight dibenzo-α-pyrones present in shilajit. It has been suggested that humic and fulvic acids act as carrier molecules for delivering the entrapped bioactive molecules present in shilajit to their intended site of action in the body. Such entrapment is also capable of enhancing the stability of the drug molecules. In fact, it has been reported that the bioactive principles of shilajit owe their stability in the natural habitat due to their entrapment in the voids (micropores) of the fulvic acids of shilajit humus (Ghosal *et al.*, 1991). The antioxidant and rejuvenating properties of shilajit could also be due to the trapping of free radicals and toxins from the body by fulvic and humic acids.

The chemical structure of these two acids suggest that these have structure exhibiting a hydrophilic exterior and a hydrophobic interior somewhat similar to that observed in case of cyclodextrins. The hydrophobic interior of these molecules is thus capable of forming complexes with nonpolar solutes and drug molecules with low bioavailability. These drug molecules can be entrapped or complexed in the hydrophobic interior so as to increase their solubility and dissolution rate, thereby enhancing their bioavailability. It was thought worthwhile by the authors and cowokers to explore and evaluate the drug complexing, dissolution, and solubility-enhancing, and surface-active properties of fulvic and humic acids extracted from shilajit and also see if such an association/interaction between the drug and fulvic and humic acids can lead to an increase in the drug bioavailability and a better pharmacodynamic profile. Research work by the authors and coworkers (Saluja, 2001; Sawnani, 2002; Khanna, 2005; Anwer, 2005, 2008; Ahmad, 2006, 2008; Tyagi, 2006; Karmarkar, 2007; Vashisht, 2007; Mirza, 2007) has proved this to be true.

There are a number of drugs in modern system of medicine which have problems of low aqueous solubility and/or low intestinal permeability which often limits their bioavailability. The Biopharmaceutical Classification System (BCS) has provided a scientific basis for classifying drugs into four classes based on their solubility and intestinal permeability as high permeability/ high solubility (class I), high permeability/low solubility (class

II), low permeability/high solubility (class III), and low permeability/low solubility (class IV) (Amidon *et al.*, 1995; Blume and Schug, 1999). Drugs belonging to the classes II, III, and IV usually present a significant challenge to the pharmaceutical industry to formulate them into highly bioavailable and therapeutically useful compositions. Although a number of approaches have been used in the past to overcome the problems of low solubility and/or low permeability, the use of these approaches have often been limited by several practical factors such as toxicity, irritancy, non-selectivity, poor drug stability, excessive size due to the need for large amount of excipients in relation to the dose, technical manufacturing problems, and the high cost of goods. Thus, there exist a definite need for identifying newer and novel approaches to overcome the problems of solubility and permeability of problematic drug candidates so as to augment their bioavailability.

The use of fulvic acids and humic acids extracted from shilajit as complexing agents for such problematic drugs can be a potential approach for increasing the bioavailability of such drugs. Although some preliminary research in this area has shown that humic and fulvic acids indeed act as carrier molecules for such drugs, more work is required before the full potential of this novel carrier can be exploited.

THE COMPLEXATION PHENOMENON

A "complex" of a species is formed by interaction of two or more molecules or ions. The following definitions are relevant in this context.

A "substrate" S is the interactant whose physical or chemical properties are observed experimentally.

A "Ligand" L is the second interactant whose concentration may be varied independently in an experimental study.

A complex is a species with a definite substrate to ligand stoichiometry, which can be formed in an equilibrium process, in solution, and also may exist in solid state.

mS	+	nL	=	Sm Ln
Substrate		Ligand		Complex

Types of Complexes

The definition of a complex leads to a classification into two groups based on chemical bonding.

(i) **Co-ordination complexes:** These are formed by coordinate bonds in which transfer of a pair of electrons takes place, *e.g.*, metal ion coordination complexes between metal ions and bases.

(ii) **Molecular complexes:** These are formed by noncovalent interactions between the substrate and the ligand such as electrostatic induction and dispersion interactions.

The molecular complexes may be classified according to the:

(a) Type of bonding or interaction, *e.g.*, charge transfer and hydrogen bonding complexes.

(B) Type of structure of interaction: enzyme substrate complex, drug-receptor complex.

(C) Type of structure of complex: self-associated aggregate, micelles, inclusion complexes, and clathrates.

Self-association: Complexation of a molecule with others of its own species, *e.g.*, benzene forms dimers.

Micelle: A special form of self-aggregated complex in which interactant is a surfactant.

Clathrate: Host molecule forms a crystal lattice containing spaces into which guest molecules can fit.

THE INCLUSION COMPLEX

An inclusion complex is formed when a macrocyclic compound possessing an intramolecular cavity of molecular dimension, interacts with a small molecule that can enter the cavity (Loftsson and Brewster, 1996). The macrocyclic molecule is called the "host" and the small included molecule is called the "guest". Synthetic macrocyclic hosts are exemplified by crown ethers, while natural macrocyclic hosts are exemplified by cyclodextrins (cycloamyloses or Schardinger dextrins) which are oligosaccharides formed by action of certain enzymes on starch. They consist of α-dglucopyranose units joined with (α-1,4)-glycosidic (ether) linkages and are water-soluble polymers (Baboota *et al.*, 2001). The binding of "guest" molecules within the "host" cyclodextrin is not fixed or permanent but rather is submitted to a dynamic equilibrium thereby affording an ease of assembly and disassembly. Binding strength depends on how well the "host-guest" complex fits together and on specific local interactions between surface atoms (Baboota *et al.*, 2003).

Complexes can be formed either in solution or in the solid state and while water is typically the solvent of choice, inclusion complexation can be accomplished in cosolvent systems and with some nonaqueous solvents. Since most of the work reported in the literature regarding inclusion complexation relates to cyclodextrins, the following information is more applicable to complexation of drugs with cyclodextrins.

FACTORS AFFECTING COMPLEXATION

A number of factors have been known to affect complexation of drugs by cyclodextrins:

Molecular Dimension of the Drug

Cyclodextrins can form complexes with drugs which have a size that is compatible with the dimensions of the cyclodextrin cavity.

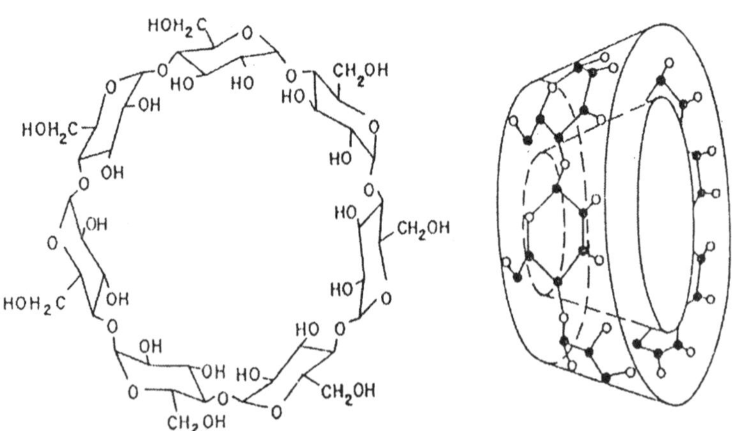

Fig. 7.1: Geometry and structure of cyclodextrin molecule

Complexation of larger molecules is also possible but in such cases only certain groups or side chains of the drug penetrate the cyclodextrin cavity. Derivatization of the hydroxyl groups of natural cyclodextrins for preparation of water-soluble derivative also imparts certain steric requirements on the part of the drug molecule to be able to form complexes. The geometry and structure of cyclodextrins is shown in Figure 7.1.

Charge and Charge Density

The charge and charge density on the cyclodextrin molecules exerts a profound impact on the complexing ability. Ionic cyclodextrins are capable of forming complexes with neutral hydrophobic drugs, if the ionic charge is not directly attached to the carbohydrate backbone of the cyclodextrin. Changing the ionization state of a drug may affect its binding to the cyclodextrin.

Temperature

Inclusion complexation is an equilibrium process and the strength of association is affected by the temperature of the system. The solubility of a drug in the cyclodextrin solution may increase with an increase in temperature even though the binding constant is decreasing, because the increase in temperature improves the solubility of the free drug (Hoshino *et al.*, 1993; Menard *et al.*, 1990).

Solvents

Organic solvents typically tend to reduce the complexation of the drug in the cyclodextrin by competing for the hydrophobic cavity (Pitha *et al.*, 1992).

Cosolubilizers

Water-soluble polymers, in low concentration, have recently been shown to increase the complexing abilities of cyclodextrins and enhance the availability of drugs in aqueous cyclodextrin solution (Loftsson *et al.*, 1998). Enhancement of complexation efficacy and increased drug availability in cyclodextrin solutions are usually obtained by heating aqueous solutions containing a water soluble polymer, cyclodextrin, and drug in an autoclave (*e.g.*, 120–140°C for 20–40 min) or by heating in a sonicator (*e.g.*, 70°C for 1 h). Simply adding the polymers to the solutions without heating does not enhance the complexation or the drug availability. In aqueous solutions, the polymers reduce the mobility of the cyclodextrin molecules and enhance the solubility of the complexes formed.

METHODS FOR EVALUATING INCLUSION COMPLEXATION

The most common and widely used method to evaluate the ability of the cyclodextrin to complex a drug is the phase solubility studies. Higuchi and Connors (1965) have classified the solubility behavior seen during complex formation (Figure 7.2) as A-type (a soluble inclusion complex is formed) or B-type (an inclusion compound of finite solubility is formed).

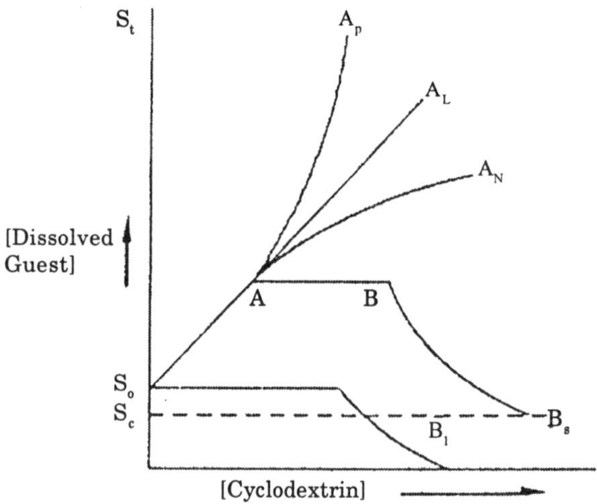

Fig. 7.2: Theoretical phase solubility diagram

The equilibrium binding or association constant (k) for the 1:1 complex can be determined from the slope of linear portion using the following relationship, where so is the intrinsic solubility of the drug under the conditions studied.

$$k_{a:b} = \frac{\text{slope}}{S_o\ (1 - \text{slope})}$$

There are a number of other methods available to determine these association or stability constants like spectroscopy [UV (Qi *et al.*, 1994), flourimetry (Duran-Meras *et al.*, 1994), NMR (Djedaini *et al.*, 1991), potentiometry (Valsami *et al.*, 1992), microcalorimetry (Tong *et al.*, 1991), freezing-point depression studies (Suzuki *et al.*, 1993), HPLC (Thuad *et al.*, 1990), and TLC techniques (Csabai *et al.*, 1993)].

METHODS FOR PREPARING COMPLEXES

Several methods have been described in literature for preparing cyclodextrin complexes of drugs. By only trial and error one can find a method, which will give the best results for a given drug. Some of these methods have been described below:

Grinding (Szejtli, 1988; Arias *et al.*, 1997)

Cyclodextrin inclusion complexes can be prepared by simply grinding the guest with cyclodextrin. This is a very slow process for making inclusion complex and degree of complexation achieved is also very low.

Solid Dispersion/Coevaporated Dispersion (Kumar *et al.*, 2003)

The drug is dissolved in ethanol and cyclodextrin is either dissolved in alcoholic solution or dissolved separately in water or other suitable medium. The cyclodextrin solution is then added to the drug solution or vice-versa and stirred to attain equilibrium. The resulting solution is evaporated to dryness preferably under vacuum.

Neutralization Method (Martin and Udupa, 1995)

Martin and Udupa (1995) reported this method for various fluoroquinolones. In this method, equimolar concentrations of drug and cyclodextrin are separately dissolved in 0.1 N NaOH, mixed and stirred for about half an hour, pH is recorded, and 0.1 N HCl is added dropwise with stirring until pH reaches 7.5, whereupon complex precipitates. The residue is filtered and washed until free from Cl-. It is dried at 25 °C for 24 hours and stored in a dessicator.

Kneading (Palmieri *et al.*, 1997)

In this method, cyclodextrin is not dissolved but kneaded like a paste, with small amount of water to which the guest component has been added. Guest component can be added without a solvent or in small amount of ethanol in which guest has been suspended. Several hours of grinding of paste in mortar results in evaporation of solvent and formation of powder-like complex.

Precipitation (Sanghavi *et al.*, 1995)

The guests which show Bs type phase solubility curve are suitable for this method of complex formation. In this method, the drug (guest) and cyclodextrin are dispersed in water and the solution is heated to obtain concentrated, viscous, and translucent liquid. The solution is left to give a precipitate of inclusion complex. Precipitate obtained is separated and dried to get solid inclusion complex.

Spray Drying (Bietti *et al.*, 1992; Piel *et al.*, 1997)

In this method, first a monophasic solution of drug and cyclodextrin is prepared using a suitable solvent (generally hydroalcoholic solutions are used). The solution is then stirred to attain equilibrium following which the solvent is removed by spray drying.

Freeze Drying (Becirevic-Lacan *et al.*, 1996; Singh and Agarwal, 2002)

Freeze-drying method is similar to spray-drying method except that in this method, after attaining the equilibrium, the solvent is removed by freeze drying.

Preparation in Suspension (Szejtli, 1988)

Cyclodextrin need not be dissolved. Simply stirring the guest in an aqueous suspension of cyclodextrin can achieve complexation within 2-24 hrs at ambient temperature. This is a recommended method for industrial application.

Melting (Szejtli, 1988)

Complexes can be prepared by simply melting the guest, mixed with finely powdered cyclodextrin. In such cases there has to be large excess of guest, and after cooling this excess is removed by careful washing with a weak complex-forming solvent or by vacuum sublimation. The latter is preferred method and is used to sublimate guests such as menthol.

CHARACTERIZATION OF COMPLEXES

The formation of inclusion complex can be studied and characterized in two ways (Szejtli, 1988).

Characterization in Solid State

Differential Scanning Calorimetry (DSC)

DSC is the measurement of rate of heat evolved or absorbed by the sample during a temperature program. The DSC curve of β-cyclodextrin generally shows an endotherm near 100 °C which signifies removal of water. The DSC curve of the guest molecule shows a sharp intense peak (endotherm) at its melting temperature (m.p.) and when it starts decomposing. In DSC curve of cyclodextrin-guest inclusion complex, these peaks are either diminished or absent. Partial complex formation may be shown by varying patterns, *e.g.*, small exotherm adjacent to the melting endotherm of guest molecule. The DSC curve of simple mixture would resemble, the combination of curves of pure substances, *i.e.*, guest and cyclodextrin.

Powder X-ray Diffraction (XRD)

This is an important technique for determination of three-dimensional structure of molecule and distinguishing between amorphous and crystalline forms. The diffraction pattern is characteristic of a substance. The crystalline substance has sharp intense peaks in its powder diffraction pattern whereas amorphous substance shows only undefined, broad, diffused peaks of low intensities. The pure drug in its free form is represented by sharp, intense peaks (crystalline nature) whereas complex has an amorphous nature, *i.e.*, broad, undefined peaks with low intensities.

Fourier Transform Infra-Red Spectroscopy (FTIR)

It is another useful technique to verify complex formation. The guest molecules within the cyclodextrin cavity show shifts in its peaks or show peaks of less intensity. Basically, peaks, which lie in the fingerprint region, and peaks due to C-O or O-H stretching are affected (shifted or intensity is changed). FTIR technique is known to have superior sensitivity and resolution, absolute wavelength accuracy, and higher precision of measurement than conventional IR technique.

Scanning Electron Microscopy (SEM)

SEM is done to observe crystal structure of the samples. SEM studies help to observe changes that occur in crystal structure during or after the preparation procedure.

Characterization in Solution

Solubility Studies and Dissolution Tests

An increase in solubility of a potential guest with increasing cyclodextrin concentration indicates complex formation in solution. When the assumed complex is dispersed in water, a very rapid dissolution, and in most cases an enhanced solubility is observed.

Thin Layer Chromatography (TLC)

TLC may also be useful for verification of complex formation, since the Rf values are altered considerably. Rf values are usually diminished provided the complex is sufficiently stable in solvent mixture used.

Proton Nuclear Magnetic Resonance (^{1}H NMR)

It is useful not only for verification of complex formation but also to guess how the guest is geometrically aligned in the cyclodextrin cavity. The inclusion of a guest molecule into the cyclodextrin cavity clearly induces some changes in the chemical shift values. The chemical shift values are also indicative of the interactions, if any, between protons of cyclodextrin and guest (Piel *et al.*, 1997; Moyano *et al.*, 1997b).

UV Studies

A constant amount of host is added to increasing concentration of cyclodextrin. UV spectra are taken and absorbance is recorded at λmax. The UV spectra shifts are attributed to a partial shielding of the excitable drug electrons into the β-cyclodextrin cavity. For example, a bathochromic shift along with a decrease in absorption on addition of β-cyclodextrin may refer to complex formation (Moyano *et al.*, 1997a).

MODEL DRUG CANDIDATES FOR STUDYING COMPLEXATION WITH HUMIC SUBSTANCES

A number of drug substances including itraconazole, ketoconazole, furosemide, melatinin, meloxicam, and piroxican possessing poor solubility and permeability characteristic were choosen as model drug candidates for studying the complexation phenomenon with humic substances derived from shilajit.

I. ITRACONAZOLE

Itraconazole (Figure 7.3) is a broad spectrum, triazole antifungal drug possessing very poor aqueous solubility and a poor bioavailability. It is a white or almost white powder, odorless and has a bitter taste. It is optically active, the racemate being used clinically. It is generally classified as a BCS Class IV drug. Currently, itraconazole is available as 100 mg capsule, 10 mg/mL oral solution and 10 mg/mL injection. The capsule that is available is prepared by a lengthy and tedious process which involves coating of a solution of itraconazole and hydroxypropyl methylcellulose on sugar beads followed by a coating of polyethylene glycol 20000. Even this formulation results in poor

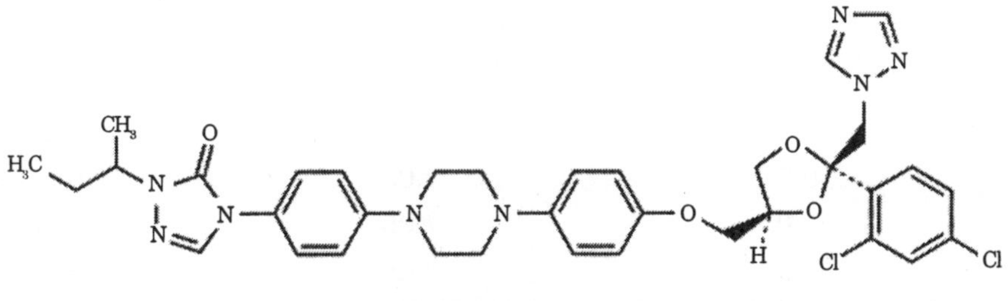

Fig. 7.3: Itraconazole

and highly variable bioavailability of itraconazole. The absolute bioavailability of itraconazole by the oral route has been determined to be only 55%. An oral solution and injectable preparations have been formulated by complexing the drug with hydroxypropyl-β-cyclodextrin. Although complexation of itraconazole with hydroxypropyl-β-cyclodextrin has resulted in some increase in its bioavailability, the quantity of hydroxypropyl-β-cyclodextrin required to solubilize itraconazole is so large that it precludes formulation of a solid dosage form using this approach.

PREPARATION OF ITRACONAZOLE COMPLEXES

Khanna (2005) studied the complexation behavior of itraconazole with fulvic acid extracted from shilajit. Various techniques such as freeze drying, solvent evaporation, and spray drying were used for the preparation of complexes. The complexes were prepared in the molar ratio of 1:0.5, 1:1, and 1:2 of drug:complexing agent by the solvent evaporation technique and since 1:1 molar ratio gave the best results, complexes by the other techniques were prepared only in 1:1 molar ratio. The quantity of itraconazole and fulvic acids used for the preparation of complexes in the different molar ratios is shown in Table 7.1.

Preparation of itraconazole-fulvic acid complexes by solvent evaporation in a rotary evaporator

Complexes of itraconazole and fulvic acid in the molar ratio of 1:0.5, 1:1, and 1:2 were prepared by dissolving the required

Table 7.1: Quantity of itraconazole and fulvic acid used for complexes prepared in different molar ratios

Ratio (drug:complexing agent)	Quantity of itraconazole (g)	Quantity of fulvic acid[a] (g)
1 : 0.5	7.05	6.0
1 : 1.0	7.05	12.0
1 : 2.0	7.05	24.0

[a]Average molecular weight, 1200

quantity (Table 7.1) of itraconazole in 100 ml of glacial acetic acid and fulvic acid in 150 mL of water. The fulvic acid solution was then added to the itraconazole solution with stirring and the solution was sonicated in an ultrasonicator bath for 3 hours. The solution thus obtained was dried in a rotary evaporator under vacuum to yield the itraconazole-fulvic acid complex. Evaporation was carried out at 100 °C by dipping the rotating flask in a boiling water bath. The dried complex was sieved through sieve no. 60 and stored in a vacuum desiccator till use.

Preparation of itraconazole-fulvic acid complexes by freeze drying

Freeze-dried complexes of itraconazole with fulvic acid in the ratio of 1:1 were prepared in a manner similar to above, with the difference that the solution of itraconazole-fulvic acid complex was frozen to -70 °C and then dried for 24 hours in a freeze dryer to obtain the freeze-dried complex. The dried complex was sieved through sieve no. 60 and stored in a vacuum desiccator till use.

Preparation of itraconazole-fulvic acid complex by spray drying

Spray-dried complexes of itraconazole with fulvic acid in the ratio of 1:1 were prepared in a manner similar to above, with the difference that the solution of itraconazole-fulvic acid complex was spray dried at an inlet temperature of 260–280 °C and a flow rate of 10 mL/min per minute to obtain the spray dried complex. The obtained complex was stored in a vacuum desiccator till use.

Preparation of itraconazole-fulvic acid complexes by physical mixing

Physical mixtures of itraconazole and fulvic acid were prepared by intimately mixing the required quantities of itraconazole and fulvic acid in a pestle mortar.

Preparation of itraconazole-fulvic acid complex by freeze drying the physical mixture.

A physical mixture of itraconazole and fulvic acid in 1:1 molar ratio was prepared by intimately mixing 7.05 g of itraconazole and 12.0 g of fulvic acid in a pestle mortar. The mixture was added to 250 mL of water and sonicated in an ultrasonicator for 3 hours. The resulting mixture was frozen at -70 °C and dried in a freeze drier for 24 hours. The dried complex was sieved through sieve no. 60 and stored in a vacuum desiccator till use.

CHARACTERIZATION OF COMPLEXES

The prepared itraconazole-fulvic acid complexes were characterized by means of differential scanning calorimetry (DSC), X-ray diffraction (XRD), Fourier transform infrared spectroscopy (FTIR), and scanning electron microscopy (SEM).

DIFFERENTIAL SCANNING CALORIMETRY

DSC thermograms (instrument calibrated by using Indium as a standard with melting point at 165 °C) were recorded using a Perkin-Elmer differential scanning calorimeter. All samples were treated according to the following specifications:

Atmosphere	:	Nitrogen
Heating rate	:	10 °C/min.
Temperature range	:	40–400°C
Sample size complex	:	Itraconazole or itraconazole-fulvic acid equivalent to 2 mg of itraconazole

Itraconazole-fulvic acid complexes prepared in different ratios by solvent evaporation technique

Figure 7.4 shows the DSC thermograms of itraconazole (ITRA) as such in comparison to that of fulvic acid and fulvic acid-itraconazole complexes prepared in different ratios by the solvent evaporation technique in a rotary evaporator. As shown in figure, itraconazole exhibited a sharp endotherm at 167 °C while fulvic acid (FA) did not exhibit any sharp endotherm indicating that it does not have any defined melting point. On the other hand, it exhibited an exothermic peak in the range of 250–350 °C which could be attributed to the thermal degradation of carbohydrates, dehydration of aliphatic structures, and decarboxylation of carboxylic groups (Pietro and Paola, 2004). Complex prepared in

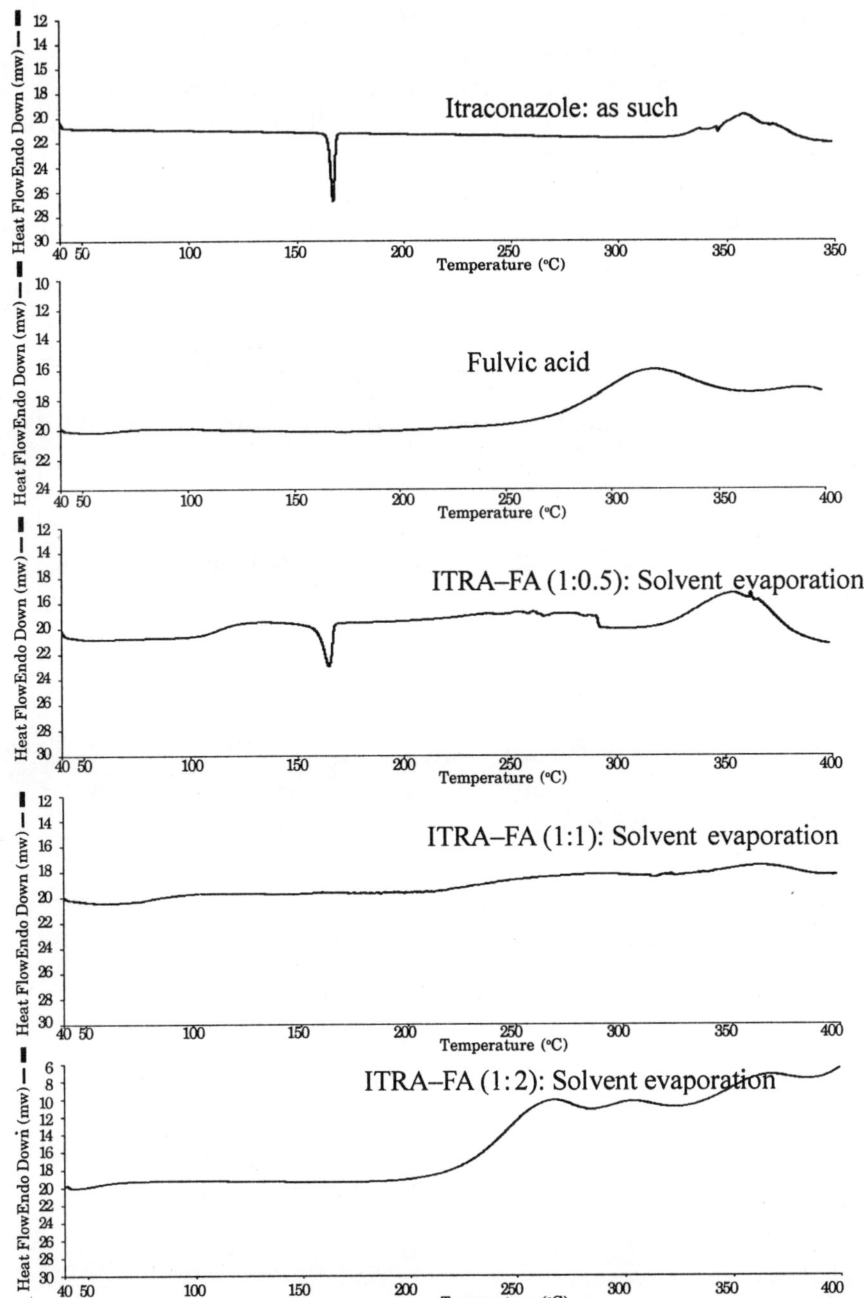

Fig. 7.4: DSC thermograms of itraconazole, fulvic acid, and itraconazole-fulvic acid complexes prepared by solvent evaporation technique

the ITRA: FA molar ratio of 1:0.5 exhibited an endotherm at around 160 °C which was also reduced in intensity. This indicates that all of itraconazole was not getting complexed due to insufficient quantity of fulvic acid and the residual itraconazole gave an endothermic peak near its melting point. DCS thermogram of complex prepared in the molar ratio of 1:1 showed complete complexation as there was neither an endothermic peak at or near the melting point of itraconazole nor an exotherm corresponding to fulvic acid. The complex in the molar ratio of 1:2 did not show an endotherm indicating that all of itraconazole got complexed but it also showed an exotherm corresponding to excess fulvic acid. Hence, it can be concluded that 1:1 was the optimum ratio for complexation between itraconazole and fulvic acid.

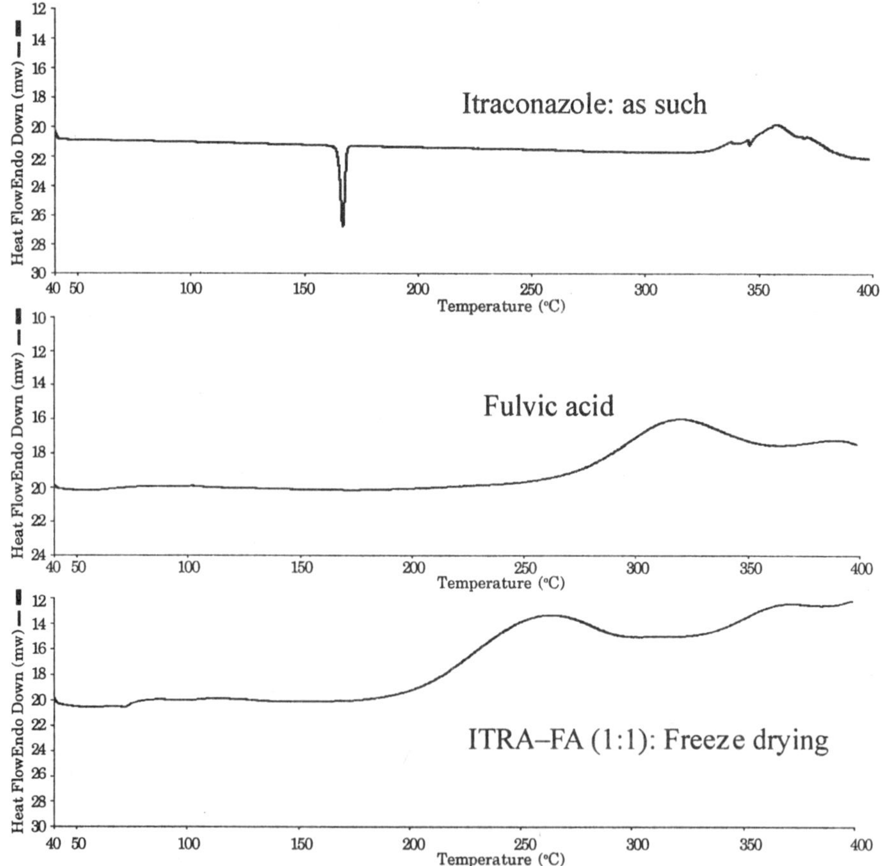

Fig. 7.5: DSC thermograms of itraconazole, fulvic acid, and itraconazole-fulvic acid complex prepared by freeze drying

Itraconazole-fulvic acid complex prepared by freeze drying technique

Figure 7.5 shows the DSC thermograms of itraconazole (ITRA) as such in comparison to that of fulvic acid and fulvic acid-itraconazole complex prepared in the molar ratio of 1:1 by freeze-drying technique. While the DSC thermogram of itraconazole exhibited a sharp endotherm at 167 °C, the peak was totally absent in the 1:1 complex prepared by freeze drying technique, demonstrating a complete complex formation between the two in this ratio.

Itraconazole-fulvic acid complex prepared by spray drying technique

Figure 7.6 shows the DSC thermograms of itraconazole (ITRA) as such in comparison to that of fulvic acid and fulvic acid-

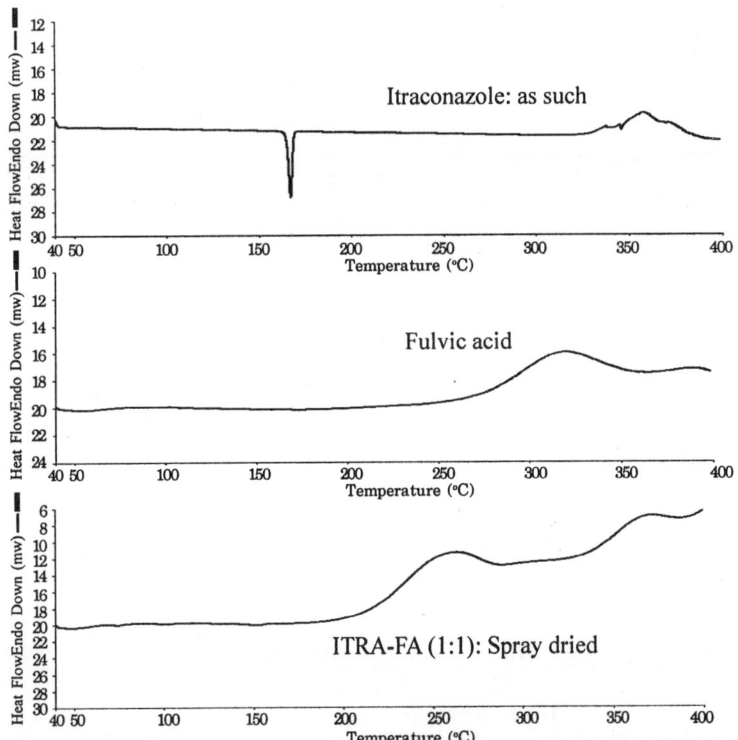

Fig. 7.6: DSC thermograms of itraconazole, fulvic acid, and itraconazole-fulvic acid complex prepared by spray drying

itraconazole complex prepared in the molar ratio of 1:1 by the spray-drying technique. While the DSC thermogram of itraconazole exhibited a sharp endotherm at 167 °C, the peak was totally absent in the 1:1 complex prepared by the spray-drying technique, demonstrating a complete complex formation between the two.

Itraconazole-fulvic acid complex prepared by physical mixing

Figure 7.7 shows the DSC thermograms of itraconazole (ITRA) as such in comparison to that of fulvic acid and fulvic acid-itraconazole complex prepared in the molar ratio of 1:1 by physical mixing. As shown in the figure, the melting endotherm of itraconazole and the exotherm of fulvic acid remained intact in the thermogram of the

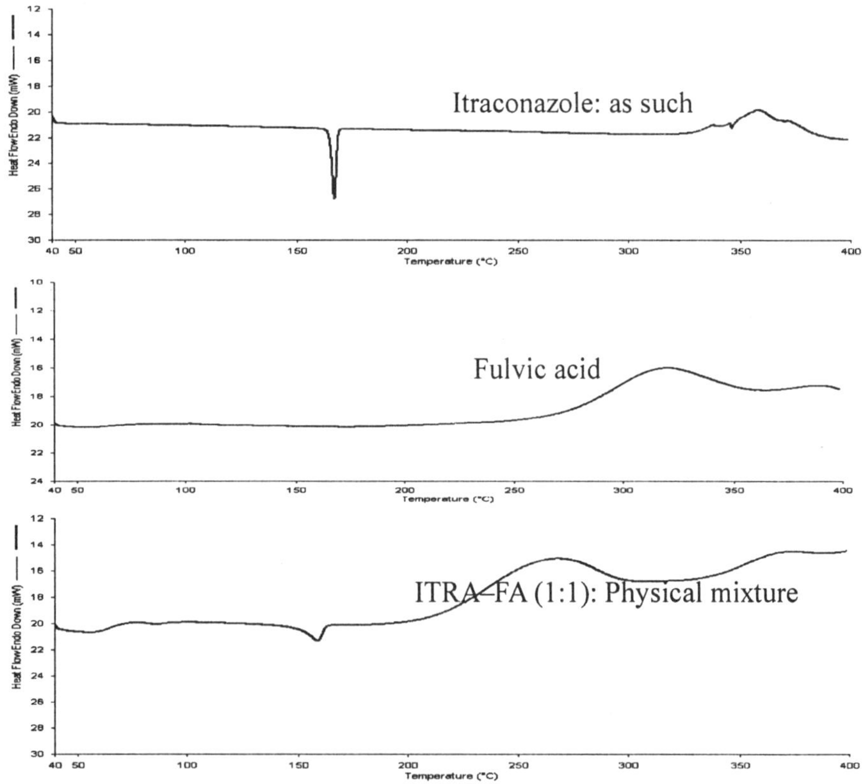

Fig. 7.7: DSC thermograms of itraconazole, fulvic acid, and itraconazole-fulvic acid complex prepared by physical mixing

complex prepared by physical mixing, demonstrating that no complex formation had taken place by the physical-mixing technique. A broadening of the itraconazole endotherm at the melting point could be attributed to the presence of fulvic acid as an impurity which would interfere with the melting endotherm of itraconazole.

Itraconazole alone

Figure 7.8 shows the DSC thermograms of itraconazole as such in comparison to thermograms for itraconazole alone that has been solvent evaporated or spray dried. As is evident from the figure, the endotherm for itraconazole remains intact even if it is dissolved in the solvent and solvent evaporated or spray dried in the absence of complexing agent fulvic acid. The intensity of

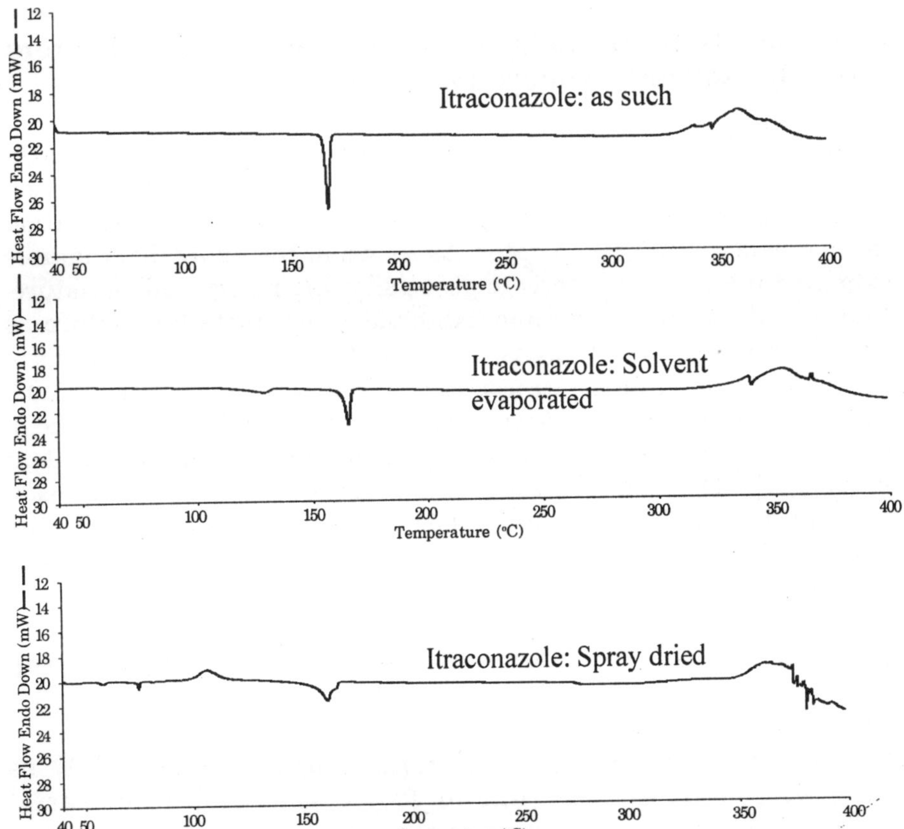

Fig. 7.8: DSC thermograms of itraconazole alone: as such, solvent evaporated, and spray dried

the peak, however, reduces most probably due to transition from the crystalline nature to the amorphous nature because of solvent evaporation or spray drying.

POWDER X-RAY DIFFRACTION

Powder X-ray diffraction patterns of the samples were obtained using a Panalytical X-ray diffractometer PW3719. All the samples were treated according to the following specifications:

Target/filter (monochromator)	:	Cu
Voltage/current	:	40 kV/50 mA
Scan speed	:	4°/min
Smoothing	:	0

Itraconazole-fulvic acid complex prepared in different ratios by solvent evaporation technique

Figure 7.9 shows the X-ray diffraction pattern of itraconazole (ITRA) as such in comparison to that of fulvic acid and fulvic acid-itraconazole complex prepared in different ratios by the solvent evaporation technique in a rotary evaporator. As shown in the figure, itraconazole exhibited intense characteristic peaks showing its crystalline nature. Fulvic acid on the other hand exhibited a noncrystalline nature as evident from the lack of intense peaks in its X-ray diffraction pattern. Complex prepared in the itraconazole: fulvic acid molar ratio of 1:0.5 exhibited a partially amorphous nature as evident by the presence of some characteristic peaks of itraconazole which were reduced in intensity. XRD patterns of complexes prepared in the molar ratios of 1:1 and 1:2 exhibited an amorphous nature characterized by the absence of intense peaks of itraconazole indicating complete complex formation between the two in these ratios.

Itraconazole-fulvic acid complex prepared by freeze drying technique

Figure 7.10 shows the XRD pattern of itraconazole (ITRA) as such in comparison to that of fulvic acid and fulvic acid-itraconazole complex prepared in the molar ratio of 1:1 by freeze-drying technique. While the XRD pattern of itraconazole exhibited sharp intense peaks, these were totally absent in the 1:1 complex

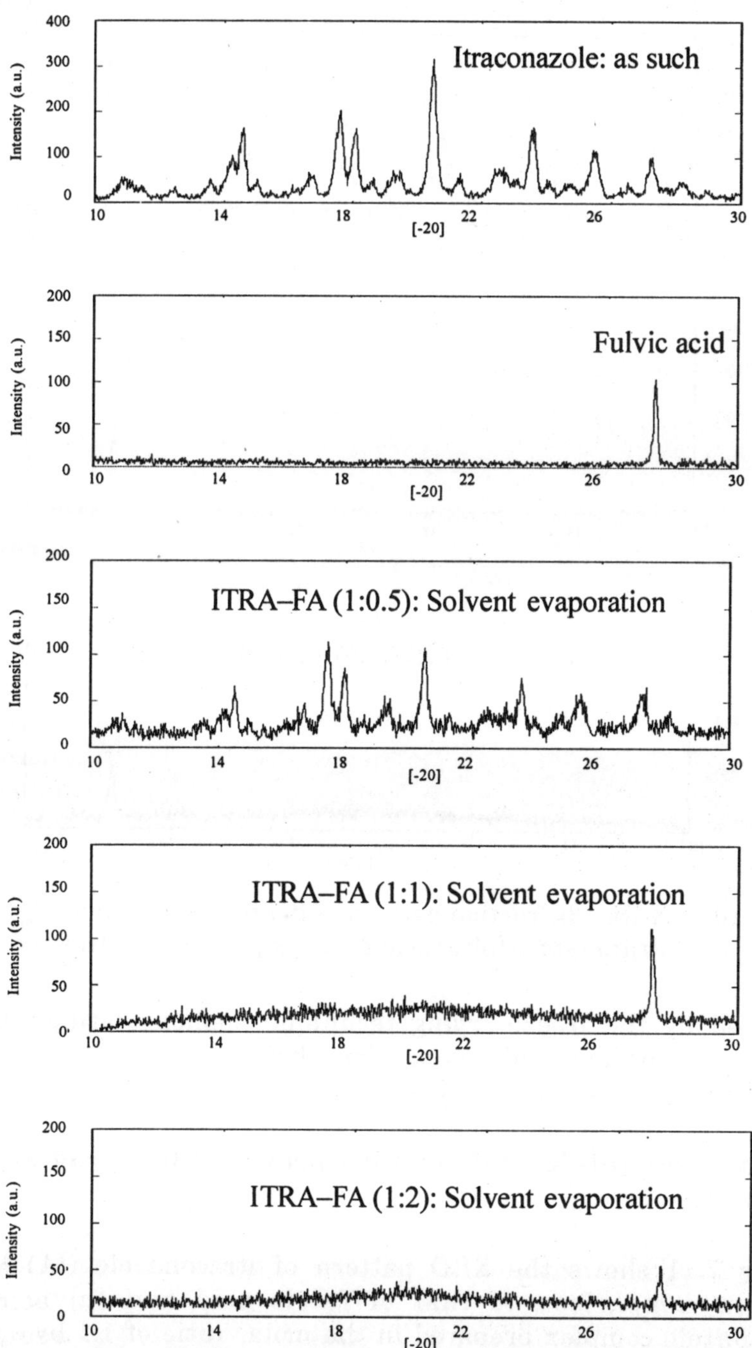

Fig. 7.9: X-Ray diffraction patterns of itraconazole, fulvic acid, and itraconazole-fulvic acid complexes prepared by solvent evaporation technique

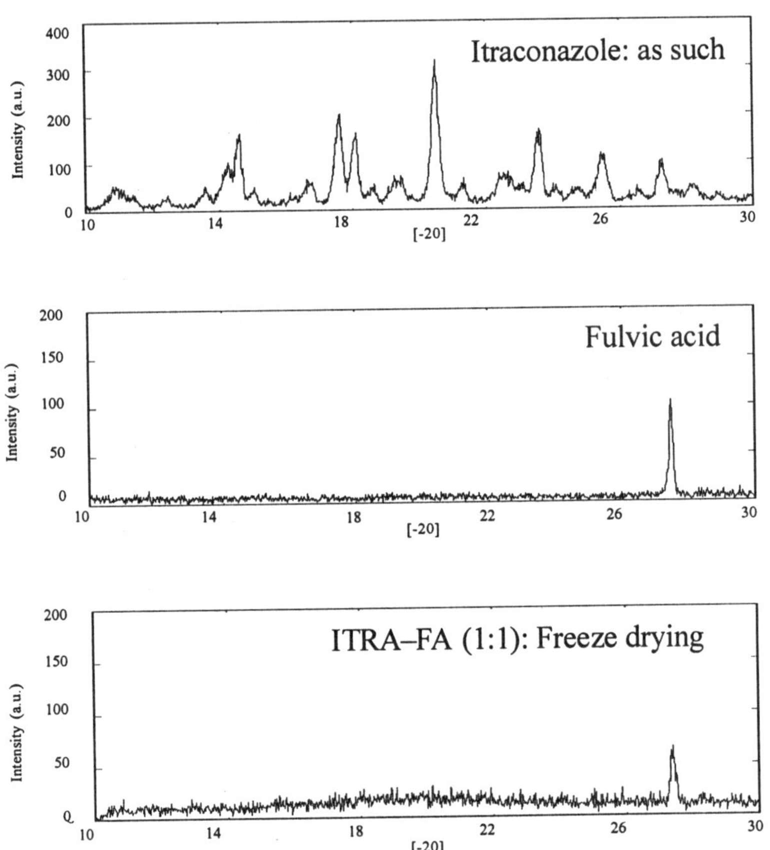

Fig. 7.10: X-Ray diffraction pattern of itraconazole, fulvic acid, and itraconazole-fulvic acid complex prepared by freeze drying

prepared by the freeze-drying technique, demonstrating that a complex formation had taken place between the two and the resulting complex was fairly amorphous in nature.

Itraconazole-fulvic acid complex prepared by spray drying technique

Figure 7.11 shows the XRD pattern of itraconazole (ITRA) as such in comparison to that of fulvic acid and fulvic acid-itraconazole complex prepared in the molar ratio of 1:1 by spray-drying technique. While the XRD pattern of itraconazole exhibited sharp intense peaks, these were totally absent in the 1:1 complex prepared by the spray-drying technique demonstrating that a

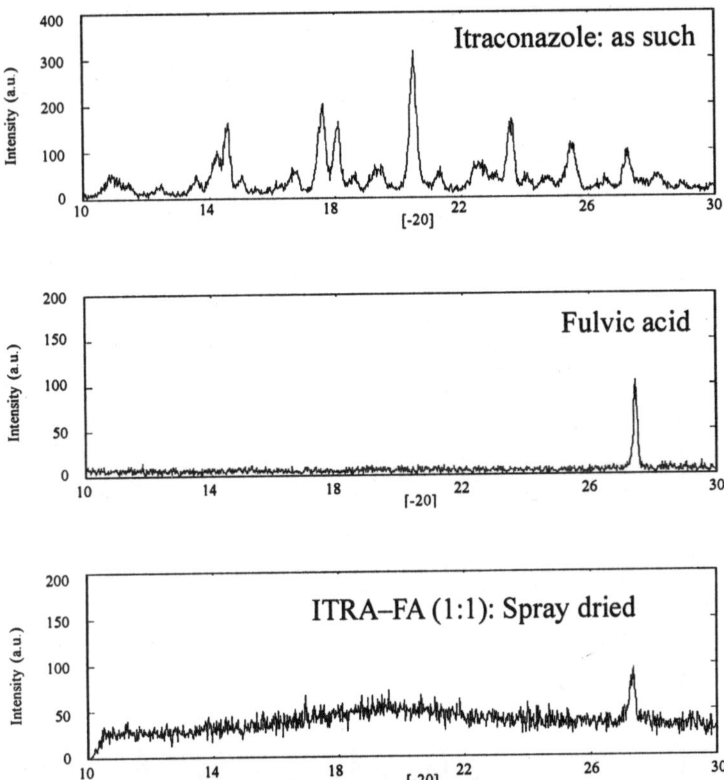

Fig. 7.11: X-Ray diffraction pattern of itraconazole, fulvic acid, and itraconazole-fulvic acid complex prepared by spray drying

complex formation had taken place between the two and the resulting complex was fairly amorphous in nature.

Itraconazole-fulvic acid complex prepared by physical mixing

Figure 7.12 shows the XRD pattern of itraconazole (ITRA) as such in comparison to that of fulvic acid and fulvic acid-itraconazole complex prepared in the molar ratio of 1:1 by physical mixing. As shown in the figure, the characteristic intense peaks of itraconazole remained intact in the complex prepared by physical-mixing technique, demonstrating that no complex formation had taken place by the physical-mixing technique and the crystalline nature of itraconazole remained unchanged.

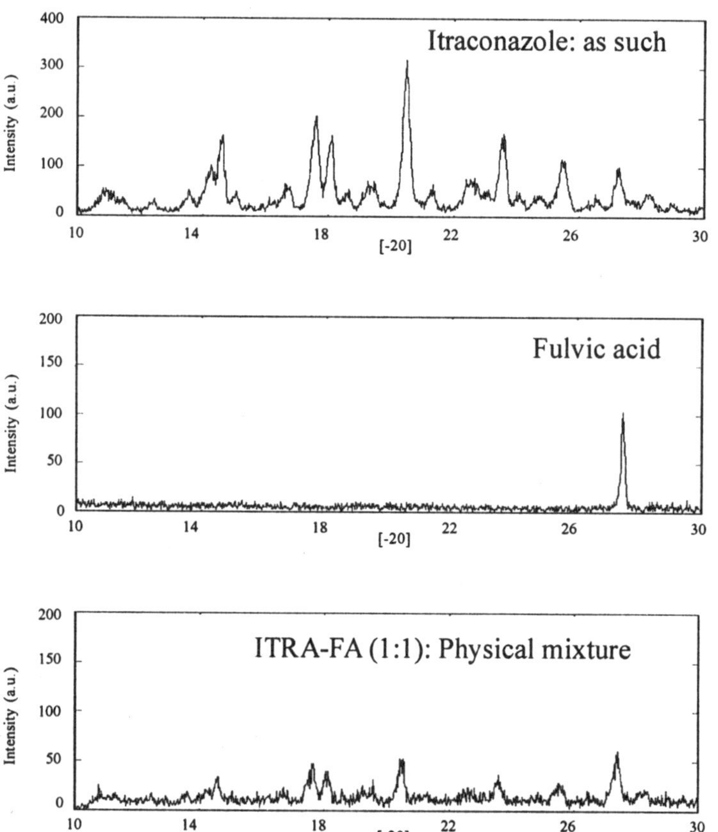

Fig. 7.12: X-Ray diffraction pattern of itraconazole, fulvic acid, and itraconazole-fulvic acid complex prepared by physical mixing

FOURIER TRANSFORM INFRA-RED SPECTROSCOPY

FTIR spectra of the samples were recorded on a Perkin-Elmer 16 PC FTIR instrument using the KBr pellet technique. Two milligram of previously dried sample was mixed with 100 mg KBr and compressed into a pellet on an IR hydraulic press. These pellets were made immediately prior to the recording of the spectrum. Scanning was done from 4000 to 450 cm^{-1}.

Itraconazole-fulvic acid complexes prepared in different ratios by solvent evaporation technique

Figure 7.13 shows the FTIR spectra of itraconazole (ITRA) as such in comparison to that of fulvic acid and fulvic acid-

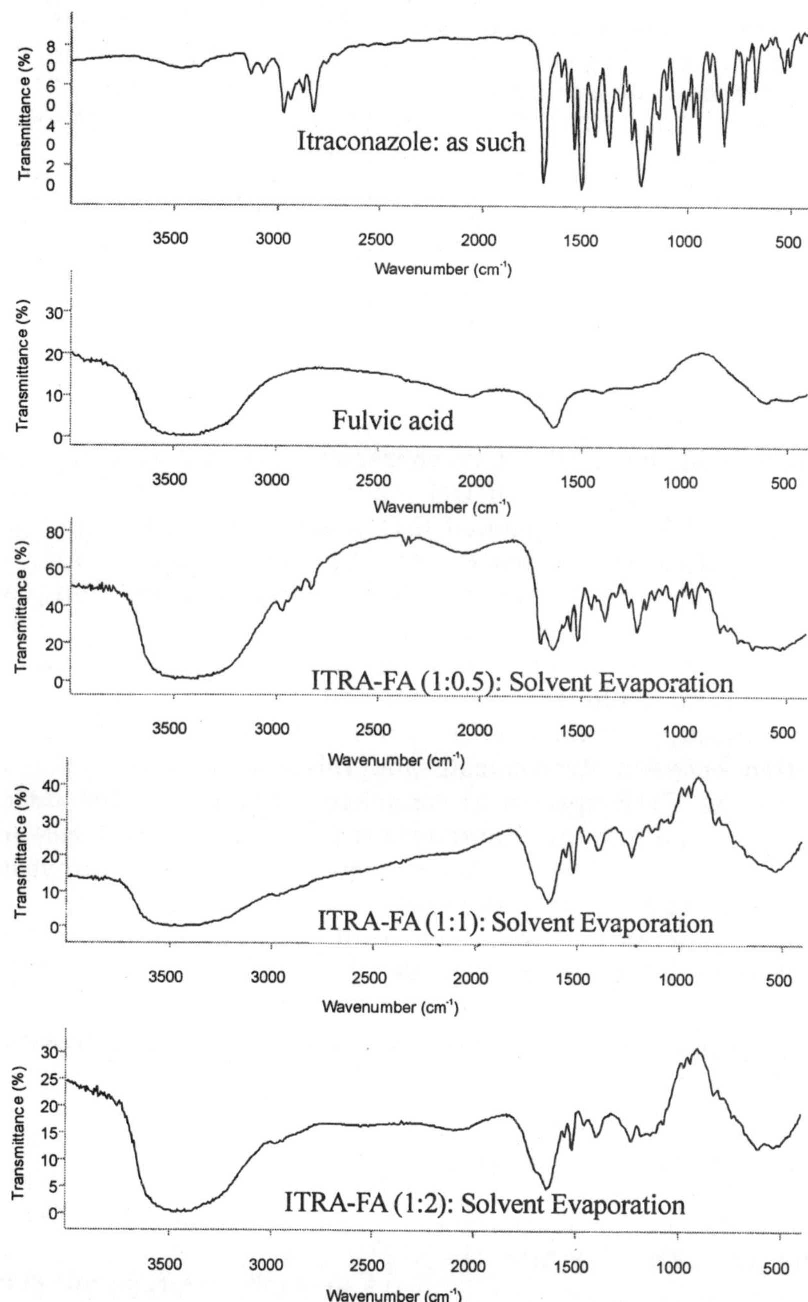

Fig. 7.13: FTIR spectra of itraconazole, fulvic acid, and itraconazole-fulvic acid complexes prepared by solvent evaporation technique

itraconazole complexes prepared in different ratios by the solvent evaporation technique in a rotary evaporator. As shown in the figure, itraconazole exhibited characteristic peaks at 3469, 3128, 3070, 2967, 1700, 1613, 1510, 1450, 1426, 1044, and 825 cm^{-1}. Bands at 3469, 3128, and 3070 could be assigned to the absorption of the NH2 groups with the first band assigned to the stretching vibration of the free NH2 group in the molecule of the pure drug and the rest of the bands at 3128 and 3070 to other amino groups (Nesseem, D.I., 2001). The bands appearing at 1613 and 1426 could be assigned to the C-N and C-N bonds, respectively, and the sharp peak appearing at 1700 could be assigned to C-O of the drug.

Fulvic acid also exhibited its characteristic broad band at about 3400 cm^{-1} (hydrogen bonded OH group) and sharp bands in the region of 1640 cm^{-1} (conjugated C-C double bond), 1400 cm^{-1} (O-H bending of carboxylic acids), and 1140 cm^{-1} (C-O stretching of polysaccharide or polysaccharide-like substances). Complex prepared in the itraconazole: fulvic acid molar ratio of 1:0.5 exhibited a spectra wherein some of the characteristic peaks of itraconazole at 3128 and 3070 cm^{-1} were missing, while some were reduced in intensity. This indicates a partial complex formation between itraconazole and fulvic acid in the ratio of 1:0.5. In the FTIR spectra of complexes prepared in the molar ratios of 1:1 and 1:2, the characteristic peaks of itraconazole were significantly reduced in number as well as intensity (3128, 3070, 2936, 2880, 2825, and 1700 cm^{-1}), indicating a good complex formation between the two which resulted in the masking of the characteristic peaks of itraconazole.

Itraconazole-fulvic acid complex prepared by freeze-drying technique

Figure 7.14 shows the FTIR spectra of itraconazole (ITRA) as such in comparison to that of fulvic acid and fulvic acid-itraconazole complex prepared in the molar ratio of 1:1 by freeze-drying technique. While the FTIR spectra of itraconazole exhibited sharp characteristic peaks at 3469, 3128, 3070, 2967, 1700, 1613, 1510, 1450, 1426, 1044, and 825 cm^{-1}, these were significantly reduced in number as well as intensity in the 1:1 complex prepared by the freeze-drying technique, demonstrating that a complex formation had taken place between the two and

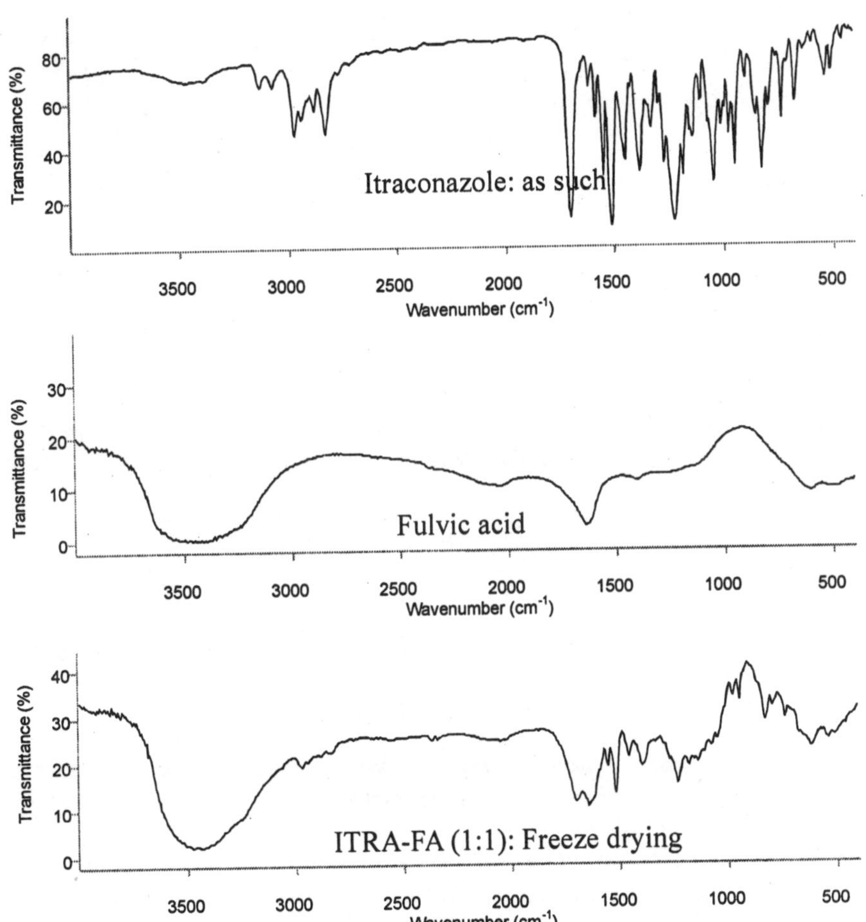

Fig. 7.14: FTIR spectra of itraconazole, fulvic acid, and itraconazole-fulvic acid complex prepared by freeze drying

this was resulting in the masking of the characteristic peaks of itraconazole.

Itraconazole-fulvic acid complex prepared by spray-drying technique

Figure 7.15 shows the FTIR spectra of itraconazole (ITRA) as such in comparison to that of fulvic acid and fulvic acid-itraconazole complex prepared in the molar ratio of 1:1 by spray-

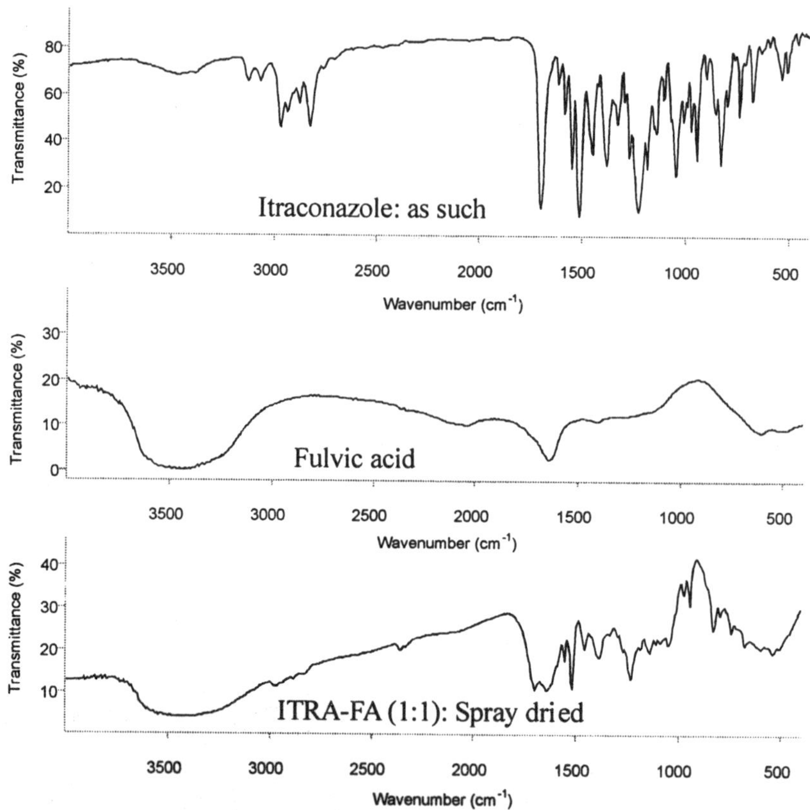

Fig. 7.15: FTIR spectra of itraconazole, fulvic acid, and itraconazole-fulvic acid complex prepared by spray drying

drying technique. While the FTIR spectra of itraconazole exhibited sharp characteristic peaks at 3469, 3128, 3070, 2967, 1700, 1613, 1510, 1450, 1426, 1044, and 825 cm^{-1}, these were significantly reduced in number as well as intensity in the 1:1 complex prepared by the freeze-drying technique, demonstrating that a complex formation had taken place between the two and this was resulting in the masking of the characteristic peaks of itraconazole.

Itraconazole-fulvic acid complex prepared by physical mixing

Figure 7.16 shows the FTIR spectra of itraconazole (ITRA) as such in comparison to that of fulvic acid and fulvic acid-itraconazole complex prepared in the molar ratio of 1:1 by physical mixing. As shown in the figure, the FTIR spectra of 1:1 complex prepared by physical-mixing exhibited peaks characteristic of both itraconazole and fulvic acid with the major peaks of itraconazole remaining intact, demonstrating that no interaction had taken place by the physical-mixing technique.

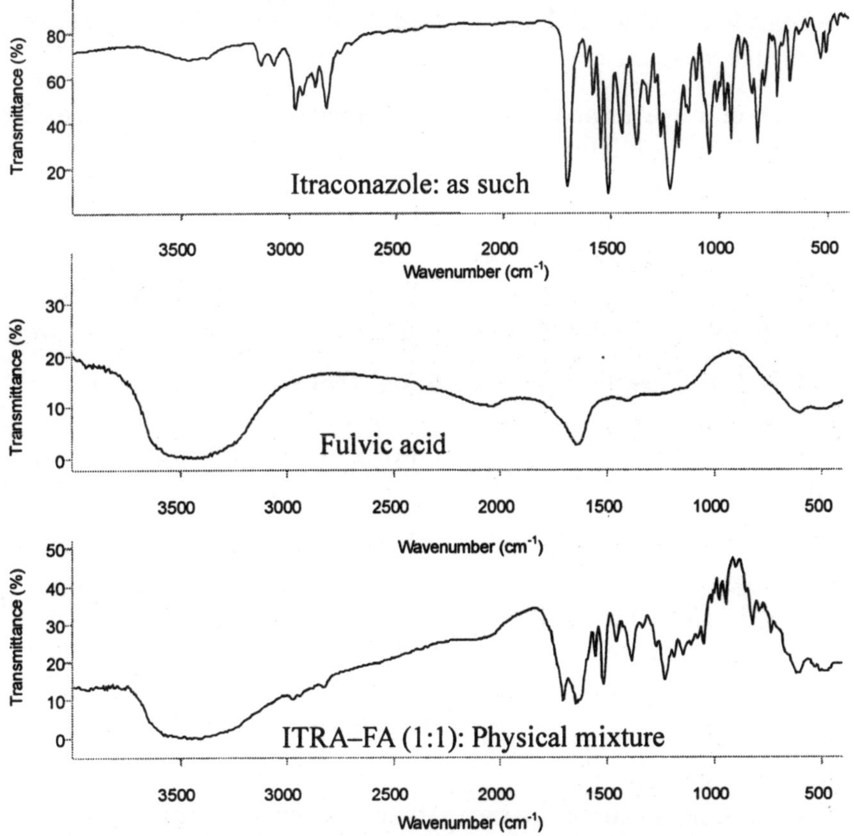

Fig. 7.16: FTIR spectra of itraconazole, fulvic acid, and itraconazole-fulvic acid complex prepared by physical mixing

SCANNING ELECTRON MICROSCOPY (SEM)

Scanning electron micrographs of prepared samples were obtained using a Joel JSM-840 Scanning Microscope with a 10kV accelerating voltage. The surface of samples for SEM were made electrically conductive in a sputtering apparatus (Fine Coat Ion Sputter JFC-1100) by evaporation of gold. Magnifications of 1500× and 3000× were used for all samples.

Figure 7.17 shows the scanning electron micrographs of itraconazole, fulvic acid, and itraconazole-fulvic acid complexes prepared in various ratios by different methods. Although this technique is not conclusive for determining the existence of a true complex in the solid state, it can be of some utility in assessing the nature and homogeneity of the solid phase.

SEM of pure itraconazole was characterized by the presence of crystalline particles of regular size, indicating its crystalline nature. Fulvic acid appeared as fibrous material. Itraconazole-fulvic acid complex prepared in a molar ratio of 1:0.5 by the solvent evaporation technique exhibited a partially crystalline nature, while those prepared in 1:1 and 1:2 ratios appeared as homogeneous amorphous mass.

The 1:1 complex prepared by freeze-drying exhibited a homogeneous, soft, and fluffy appearance. No drug crystals could be distinguished again suggesting complete complex formation.

Finally, the 1:1 spray dried complex showed the typical morphology of preparations generally obtained by this method. Small, spherical particles tending to aggregate to each other could be seen, suggesting the existence of a homogeneous product resulting from the complex formation.

On the other hand, the complex of itraconazole and fulvic acid prepared by physical-mixing technique, exhibited characteristic features of both itraconazole and fulvic acid. The crystals of itraconazole could be seen adhering to the surface of fulvic acid, indicating a lack of complex formation between the two.

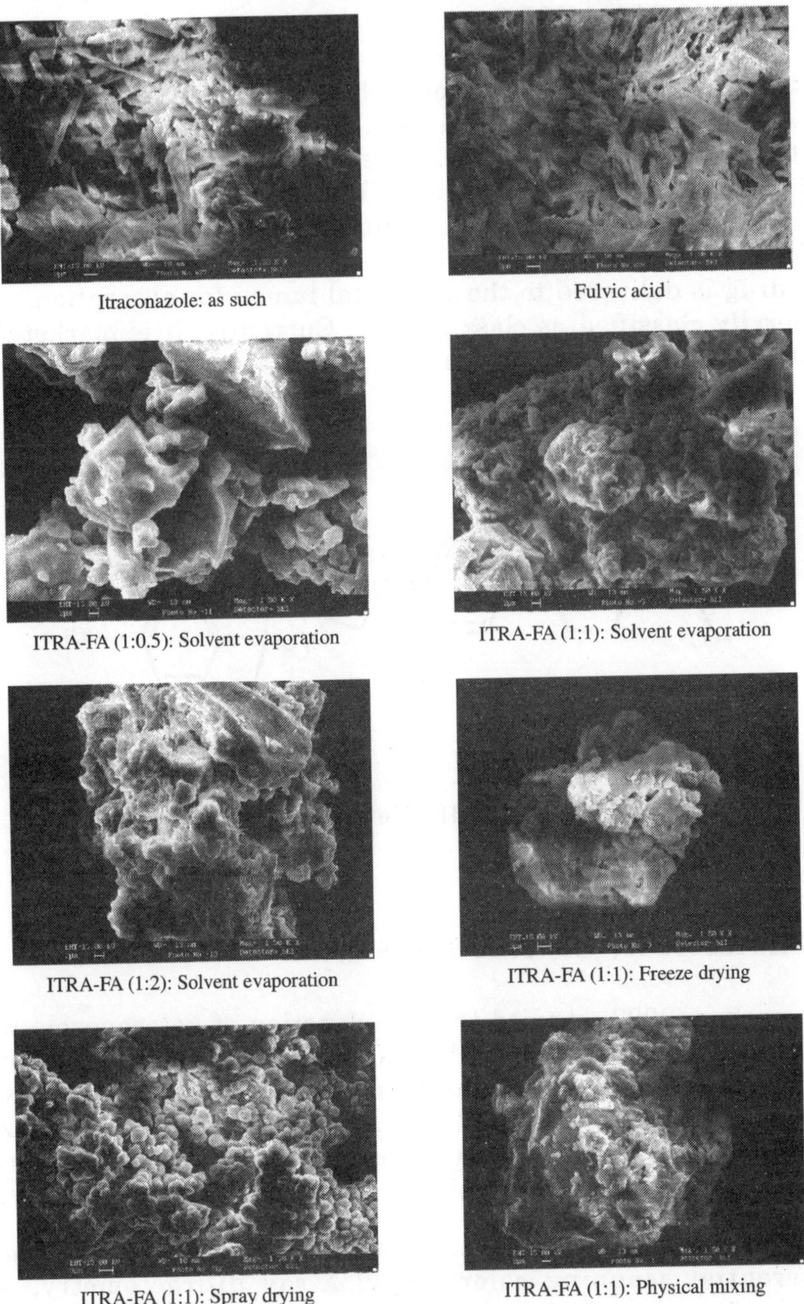

Itraconazole: as such

Fulvic acid

ITRA-FA (1:0.5): Solvent evaporation

ITRA-FA (1:1): Solvent evaporation

ITRA-FA (1:2): Solvent evaporation

ITRA-FA (1:1): Freeze drying

ITRA-FA (1:1): Spray drying

ITRA-FA (1:1): Physical mixing

Fig. 7.17: Scanning electron micrographs of itraconazole, fulvic acid and itraconazole-fulvic acid complexes prepared by different techniques

II. KETOCONAZOLE

Ketoconazole (Fig. 7.18) is another broad spectrum, imidazole class of antifungal drug having a high lipophilicity, which makes it virtually insoluble in water. It is a weak base (pKa 2.94 and 6.51) that is ionized only at low pH, such as that found in gastric fluid. One of the problems associated with the oral administration of ketoconazole is the insufficient dissolution in the stomach before the drug is delivered to the intestinal lumen for absorption. It is generally classified as class II drug. Currently, it is marketed as 200 mg tablet and as an oral suspension.

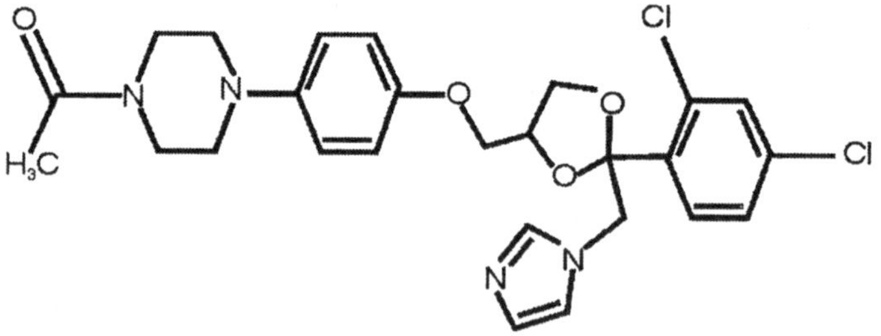

Fig. 7.18: Ketoconazole

PREPARATION OF KETOCONAZOLE COMPLEXES

Karmarkar (2007) studied the complexation of ketoconazole with fulvic acid extracted from shilajit. Complexes of ketoconazole were prepared by various techniques in different molar ratios with fulvic acid extracted from shilajit. The quantity of ketoconazole and fulvic acid used for the preparation of complexes in different molar ratio is shown in Table 7.2.

The prepared complexes were characterized by means of differential scanning calorimetry, X-ray diffractometry, and Fourier transform infrared spectroscopy techniques.

Table 7.2: Quantity of ketoconazole and fulvic acid used for complexes prepared in different molar ratios

Ratio (drug:complexing agent)	Quantity of ketoconazole (g)	Quantity of fulvic acid[a] (g)
1 : 0.5	5.31	6.08
1 : 1	5.31	12.00

[a]Average molecular weight, 1200.

DIFFERENTIAL SCANNING CALORIMETRY

DSC thermogram (instrument calibrated by using Indium as a standard having melting point at 165 °C) was recorded using a Perkin-Elmer differential scanning calorimeter equipped with computerized data station.

Ketoconazole-fulvic acid complex prepared in different molar ratio by solvent evaporation technique

Figure 7.19 shows the DSC thermogram of ketoconazole (KET) as such in comparision to that of fulvic acid and the ketoconazole-fulvic acid complexes prepared in different ratios by the solvent evaporation technique in a rotary evaporater. As shown in the figure, ketoconazole exhibited a sharp endothermic peak at 151 °C while fulvic acid (FA) did not exhibit any sharp endothermic peak indicating that it does not have any defined melting point. KET-FA complex prepared in the molar ratio of 1:0.5 exhibited an endotherm at around 150 °C which was also reduced in intensity. This indicates that all of ketoconazole is not getting complexed due to insufficient quantity of fulvic acid and the residual ketoconazole gives an endothermic peak near its melting point. DSC thermograph of complex prepared in a molar ratio of 1:1 showed complete complexation as there is neither an endothermic peak at or near the melting point of ketoconazole.

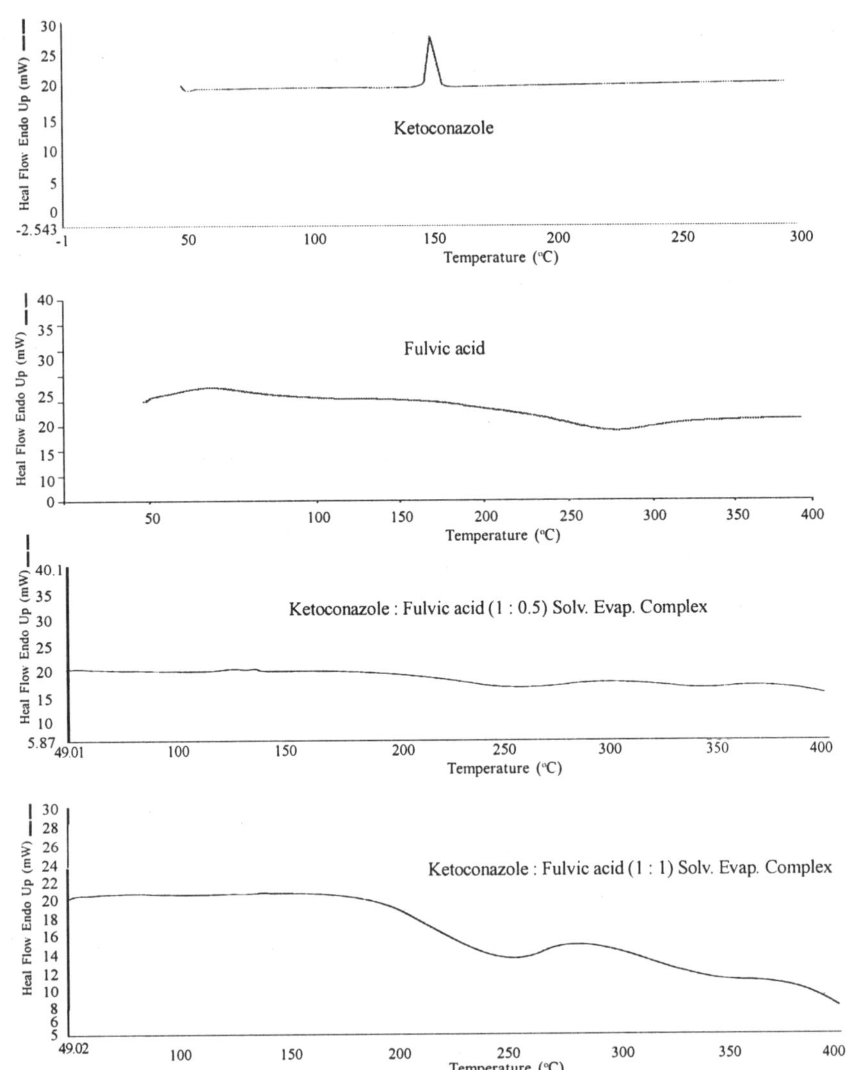

Fig. 19: DSC thermogram of ketoconazole, fulvic acid, and ketoconazole-fulvic acid complex prepared by solvent evaporation technique

Ketoconazole-fulvic acid complex prepared by spray drying technique

Figure 7.20 shows the DSC thermogram of ketoconazole (KET) as such in comparision to that of fulvic acid and ketoconazole-fulvic acid complex prepared in the molar ratio of 1:0.5 and 1:1 by the spray-drying technique.

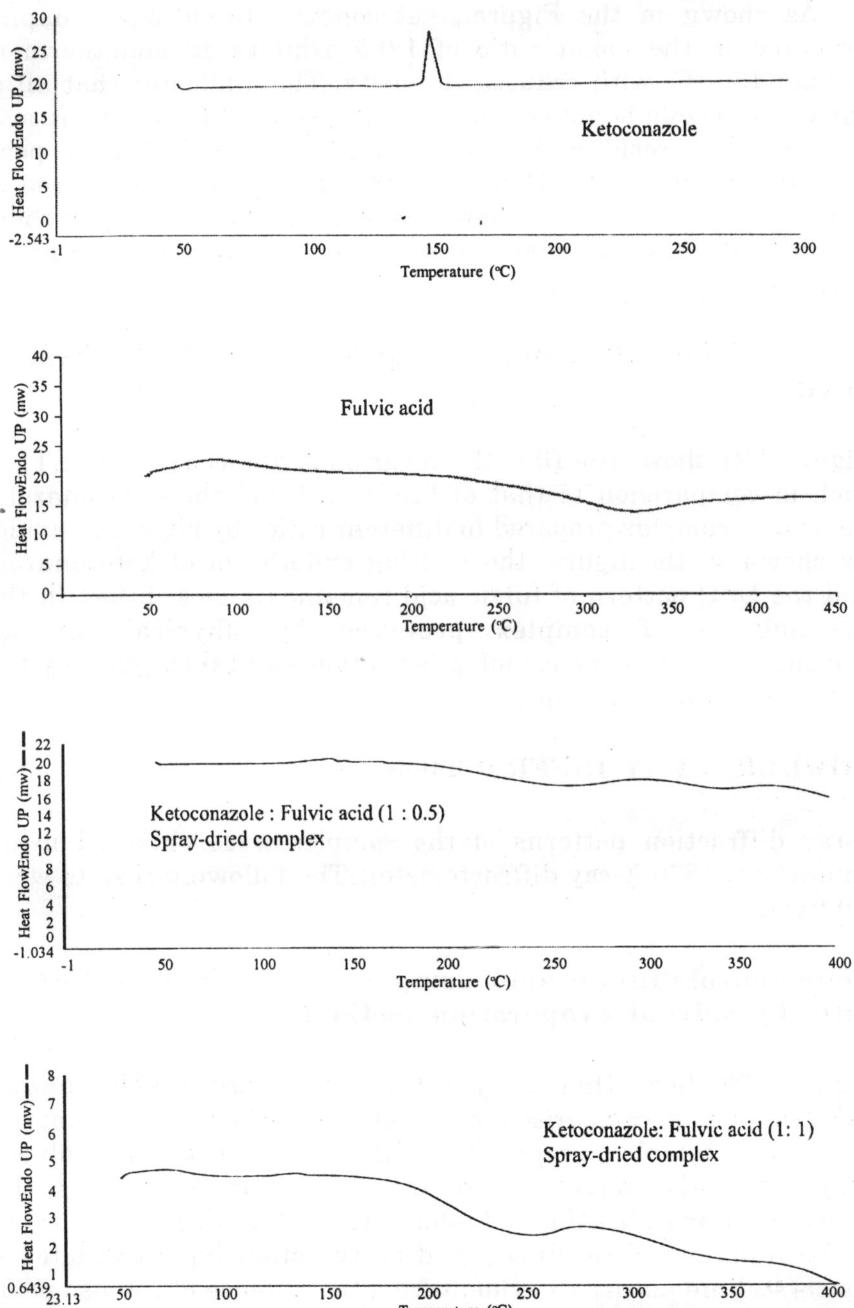

Fig. 7.20: **DSC thermogram of ketoconazole, fulvic acid, and ketoconazole-fulvic acid complex prepared by spray-drying method.**

As shown in the Figure., ketoconazole-fulvic acid complex prepared in the molar ratio of 1:0.5 exhibits an endotherm at around 150 °C, with reduced intensity. This indicates that all of the ketoconazole is not completely encapsulated in the fulvic acid pore and the residual ketoconazole gave an endothermic peak near its melting point. DSC thermogram of complex prepared in a molar ratio of 1:1 ratio showed complete complexation as there is no endothermic peak at or near the melting point of ketoconazole.

Ketoconazole-fulvic acid complex prepared by physical mixing

Figure 7.21 shows the DSC thermogram of ketoconazole (KET) as such in comparision to that of fulvic acid and the ketoconazole-fulvic acid complex prepared in different ratios by physical mixing. As shown in the figure, the melting endotherm of ketoconazole and the DSC pattern of fulvic acid remained almost intact in the thermogram of complex prepared by physical mixing, demonstrating that no complex formation had taken place by the physical-mixing technique.

POWDER X-RAY DIFFRACTION

X-ray diffraction patterns of the samples were obtained using panalytical 1830 X-ray diffractometer. The following results were obtained.

Ketoconazole-fulvic acid complex prepared in different ratio by solvent evaporation method

Figure 7.22 shows that X-ray diffraction pattern of ketoconazole (KET) as such in comparison to that of fulvic acid and ketoconazole-fulvic acid complex prepared in different ratio by the solvent evaporation technique in a rotary evaporator. As seen in the figure, ketoconazole exhibited intense characteristic peaks showing its crystalline nature. Fulvic acid on the other hand exhibited a noncrystalline nature as evident from the absence of intense peak in its X-ray diffraction pattern. Ketoconazole-fulvic acid complex prepared in the molar ratio of 1:0.5 exhibited a partially crystalline nature as evident by the lack of some characteristic peaks of ketoconazole which were of reduced intensity.

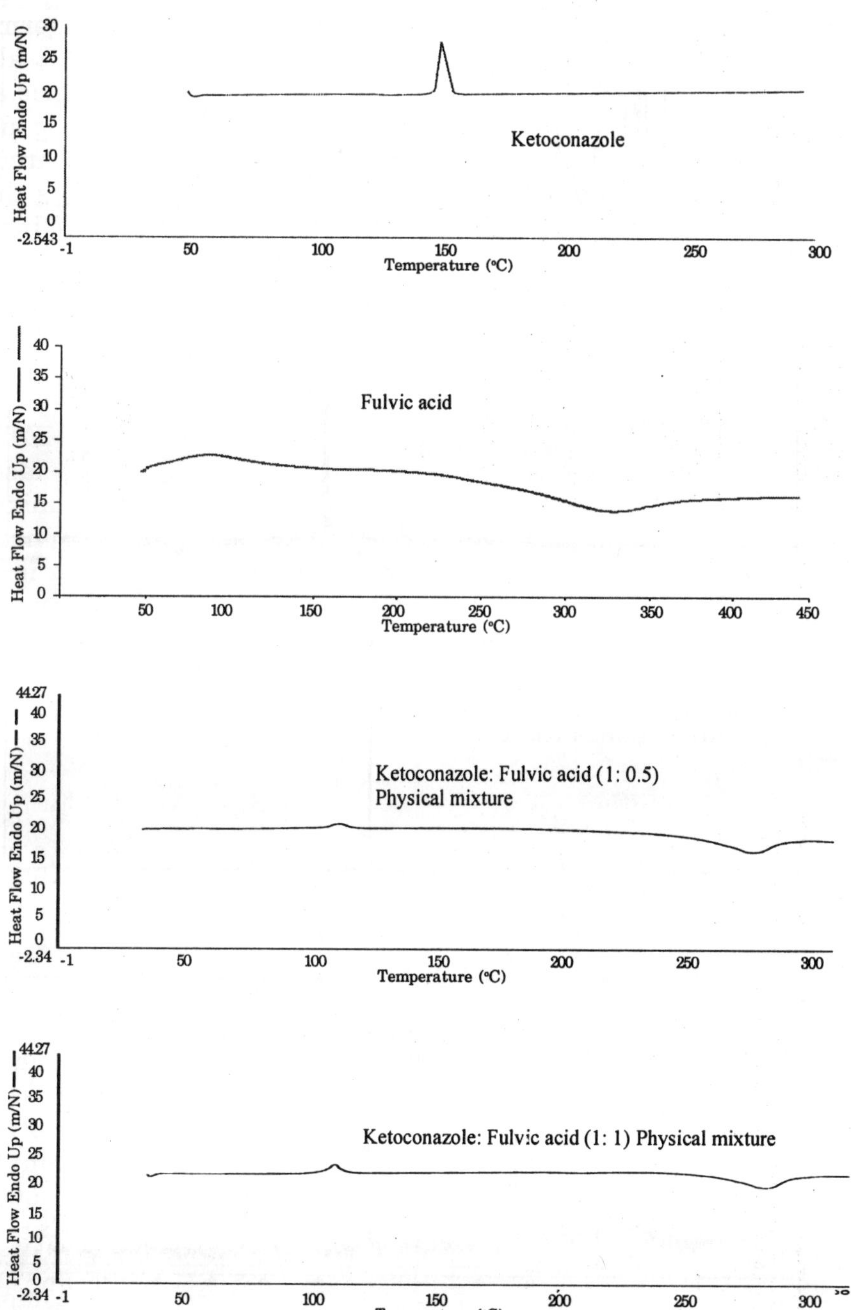

Fig. 7.21: DSC thermogram of ketoconazole, fulvic acid, and ketoconazole-fulvic acid complex prepared by physical mixing

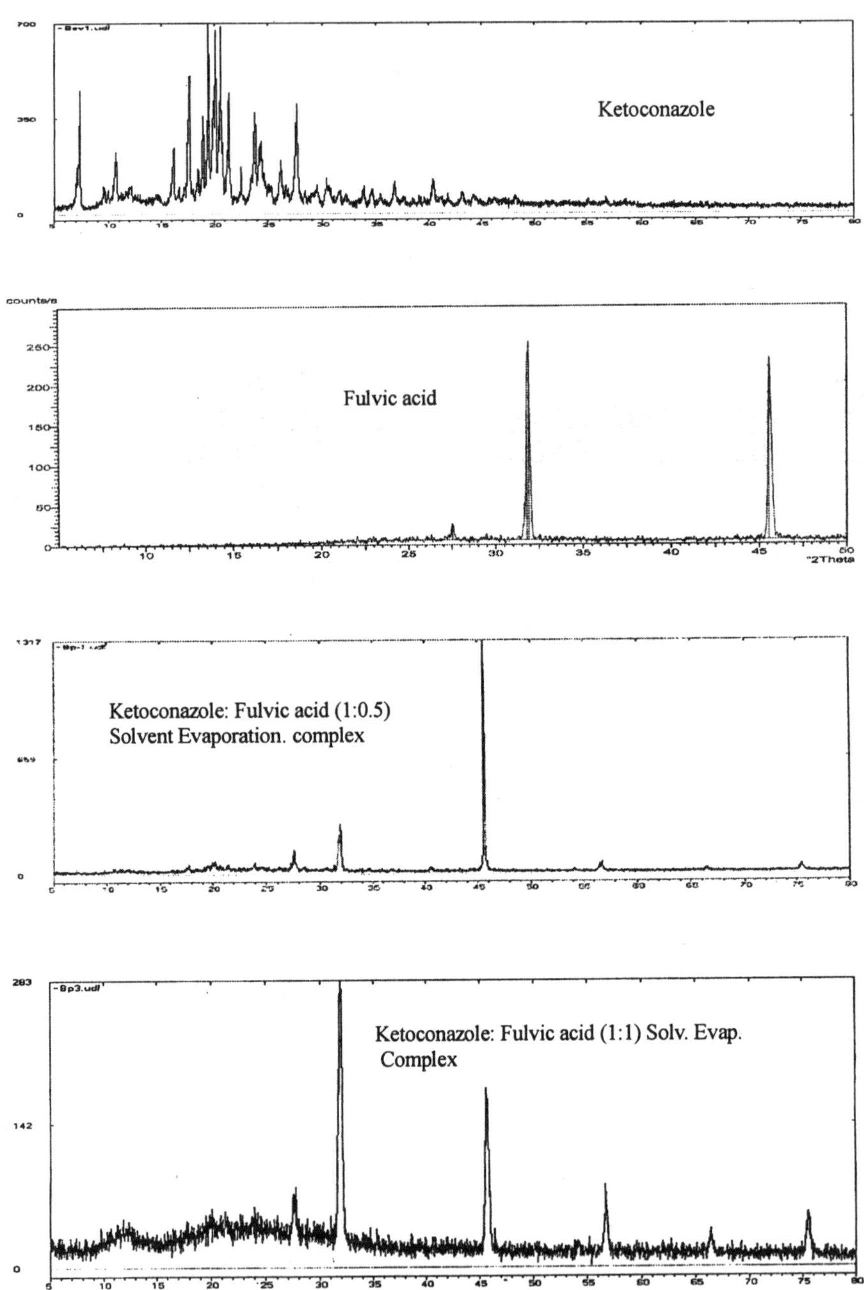

Fig 7.22: X-ray diffraction pattern of ketoconazole, fulvic acid, and ketoconazole-fulvic acid complex prepared by solvent evaporation technique

XRD pattern of complex prepared in molar ratio of 1:1 exhibited an amorphous nature characterized by the absence of intense peak of ketoconazole indicating complete complex formation between two in this ratio.

Ketoconazole-fulvic acid complex prepared by spray-drying method

Figure 7.23 shows the XRD pattern of ketoconazole (KET) as such in comparision to that of fulvic acid and ketoconazole-fulvic acid complex prepared in the molar ratio of 1.0.5 and 1:1 by spray-drying method. While the XRD pattern of ketoconazole exhibited sharp intense peak, these are totally absent in the 1:0.5 and 1:1 complex prepared by spray-drying technique

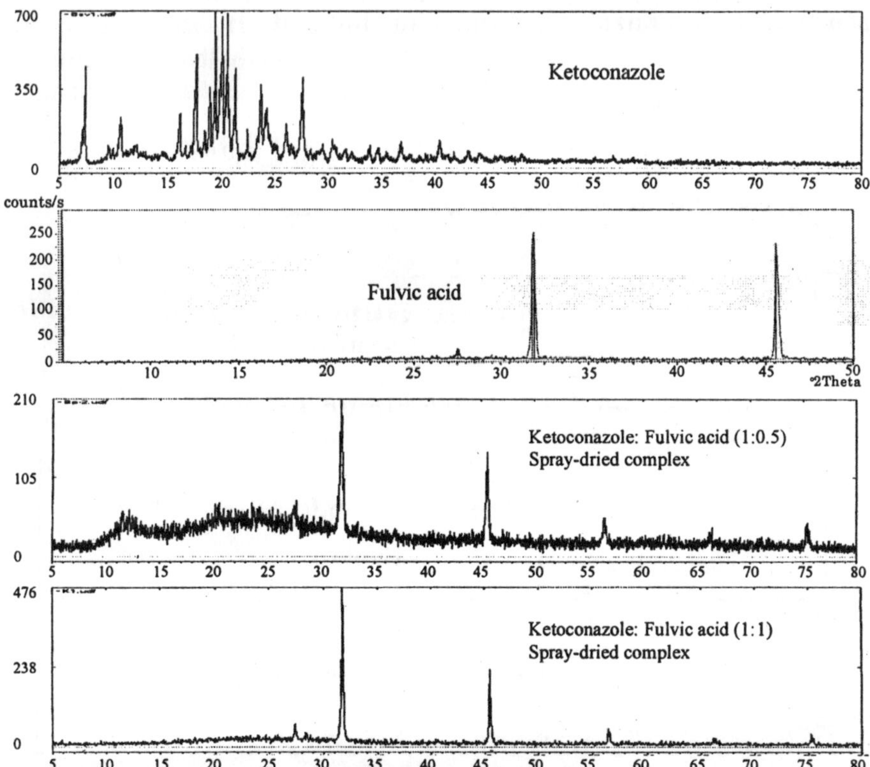

Fig 7.23: X-ray diffraction pattern of ketoconazole, fulvic acid, and ketoconazole - fulvic acid complexes prepared in different molar ratio by spray-drying method

demonstrating that complex formation has taken place between ketoconazole and fulvic acid prepared in different molar ratio by spray drying. The figure shows that complexes are fairly amorphous in nature.

Ketoconazole-fulvic acid complex prepared by physical mixing

Figure 7.25 shows the XRD pattern of ketoconazole as such in comparision to that of fulvic acid and ketoconazole-fulvic acid complex prepared in different molar ratios by physical-mixing method. As shown in the figure, the characteristic intense peaks of ketoconazole remained intact in the complex prepared by physical mixing, demonstrating that no complex formation had taken place by the physical-mixing technique and the crystalline nature of ketoconazole remained unchanged.

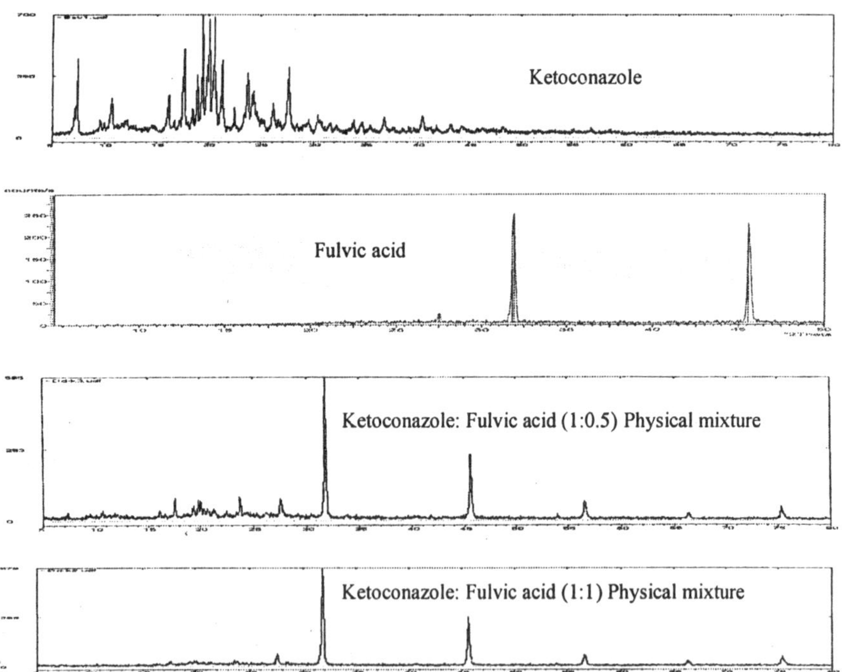

Fig 7.24: X-ray diffraction pattern of ketoconazole, fulvic acid, and ketoconazole-fulvic acid complexes prepared in different molar ratio by physical-mixing method

FOURIER TRANSFORM INFRA-RED SPECTROSCOPY (FTIR)

FTIR spectra of ketoconazole and the prepared complexes were recorded on a Perkin-Elmer 16 PC FTIR instrument using the KBr pellet technique from 4000 to 450 cm^{-1}. The following spectra were obtained.

Ketoconazole-fulvic acid complexes prepared in different molar ratio by solvent evaporation technique

Figure 7.26 shows the FTIR spectra of ketoconazole (KET) as such in comparision to that of fulvic acid and ketoconazole-fulvic acid complex prepared in different ratio by the solvent evaporation technique in a rotary evaporator. As shown in the spectra, ketoconazole exhibited characteristic peaks at 3423, 3118, 3070, 2963, 1843, 1647, 1512, 1460, 1053, and 515 cm^{-1}. Bands at 3423, 3070, and 2963 cm^{-1} could be assigned to the absorption of the NH2 group with the first band assigned to the stretching vibration of the free NH2 group in the molecule of the drug and the rest of the bands at 3070 and 2963 to other amino groups. The band appearing at 1047 and 1460 could be assigned to C=N and C-N bands, respectively, and the sharp peak appearing at 1843 could be assigned to C=0 of the drug.

Fulvic acid also exhibited its chracteristic broad band at about 3410 cm^{-1} (hydrogen bonded OH groups) and sharp band in the region of 1725 and 1635 (conjugated C=C double bond), 1406 cm^{-1} (O-H bending of carboxylic acid) and 1233 cm^{-1} (C-O stretching of polysaccharides or polysaccharides-like substances). Ketoconazole-fulvic acid complex prepared in the molar ratio of 1:0.5 exhibited spectra wherein some of the characteristic peaks of ketoconazole at 3118 and 3020 cm^{-1} were missing, while some were reduced in intensity. This indicates a partial complex formation between ketoconazole and fulvic acid in the molar ratio of 1:0.5. In the FTIR spectra of ketoconazole-fulvic acid complex prepared in the molar ratio of 1:1, the characteristic peaks of ketoconazole were significantly reduced in number as well as intensity (3118, 3070, 2963, 2882, 2830, and 1843 cm^{-1}), indicating a good complex formation between the two which results in the masking of the characteristic peaks of ketoconazole.

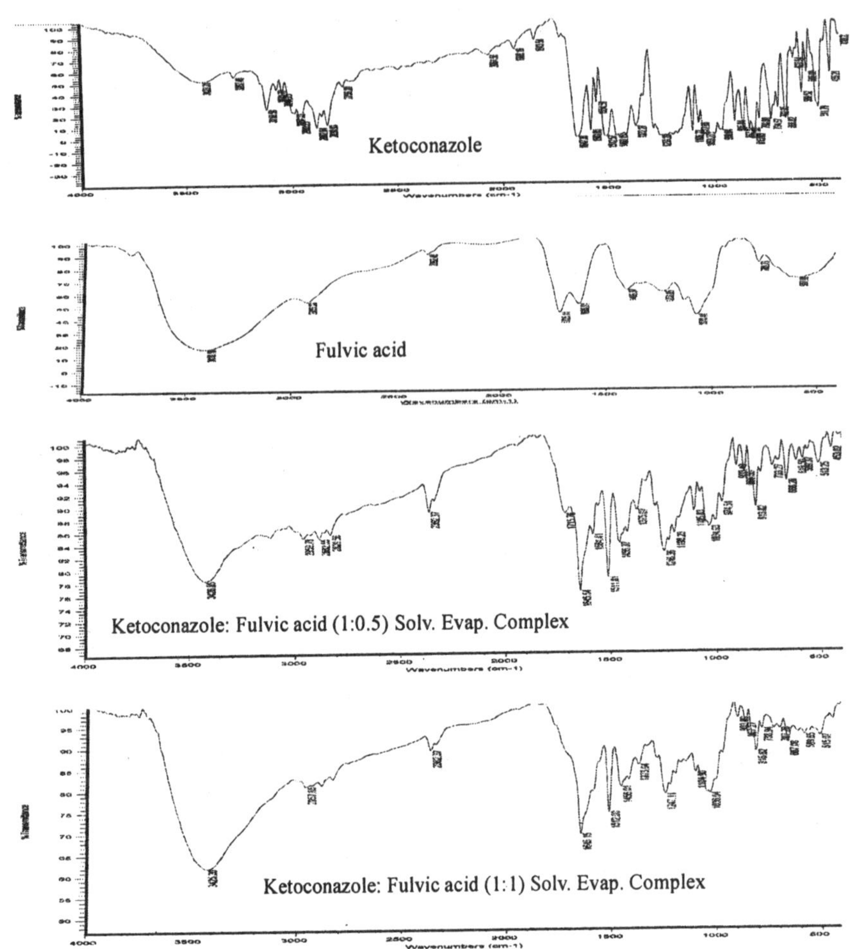

Fig. 7.25: FTIR spectra of ketoconazole, fulvic acid, ketoconazole-fulvic acid complexes prepared by solvent evaporation

Ketoconazole - fulvic acid complex prepared by spray drying technique.

Figure 7.27 shows the FTIR spectra of ketoconazole (KET) as such in comparision to that of fulvic acid and ketoconazole-fulvic acid complexes prepared in a molar ratio of 1:0.5 and 1:1 by spray-drying technique. While the FTIR spectra of ketoconazole shows characteristic peaks at 3423, 3118, 3070, 2963, 1843, 1647, 1512, 1460, 1053, and 815 cm^{-1}, these were significantly reduced in number as well as in the intensity in the 1:0.5, 1:1 complexes

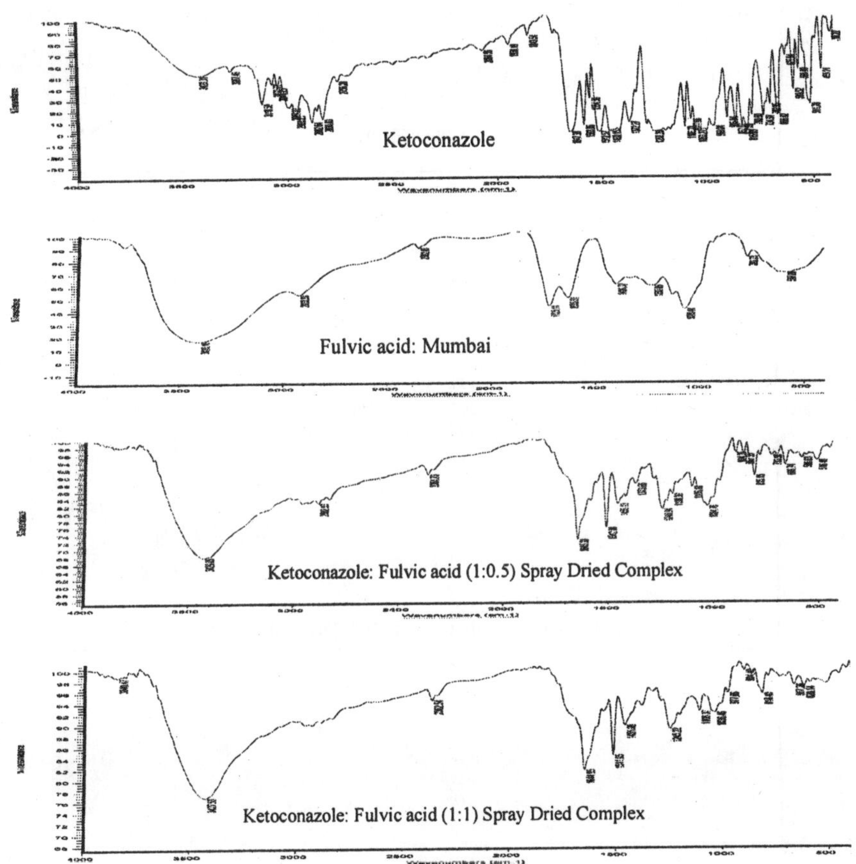

Fig. 7.26: FTIR spectra of ketoconazole, fulvic acid, ketoconazole-fulvic acid complex prepared by spray-drying technique.

prepared by the spray-drying technique demostrating that a complex formation had taken place between the two, which results in masking of the characteristic peaks of ketoconazole.

Ketoconazole-fulvic acid complex prepared by physical mixing

Figure 7.27 shows that FTIR spectra of ketoconazole (KET) as such in comparision to that of fulvic acid and ketoconazole-fulvic acid complexes prepared in a molar ratio of 1:0.5 and 1:1 by physical-mixing method. As shown in the figure, the FTIR spectra of the complexes prepared by physical-mixing method exhibited

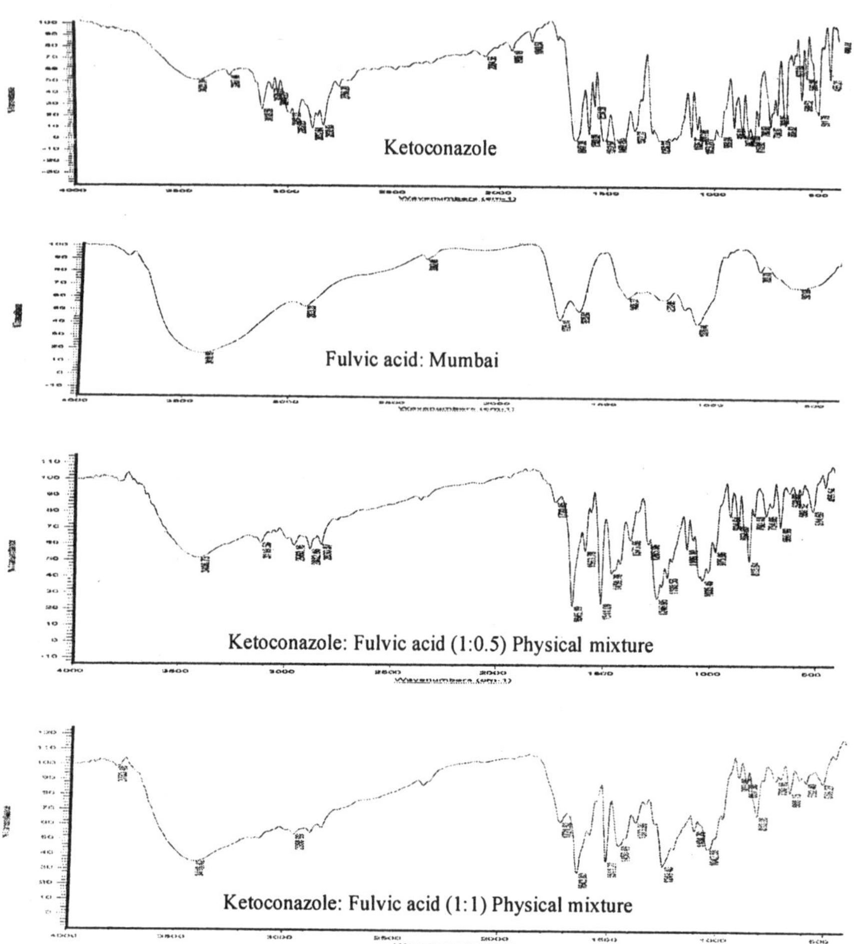

Fig 7.27: FTIR spectra of ketoconazole, fulvic acid, and ketoconazole-fulvic acid complex prepared by physical-mixing method.

peaks characteristics of both ketoconazole and fulvic acid with major peaks of ketoconazole remain intact, demonstrating that no interaction had taken place by the physical-mixing technique.

III. FUROSEMIDE

Furosemide (Fig. 7.28) is a potent loop diuretic that is used to adjust the volume and/or composition of body fluid in a variety of situations, including hypertension, heart failure, renal failure,

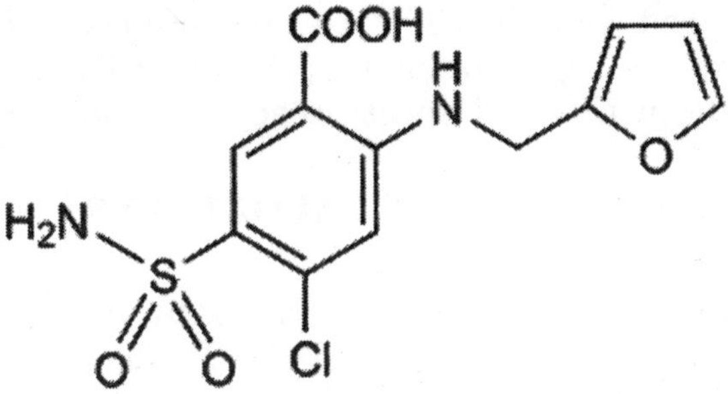

Fig. 7.28: Furosemide

nephritic syndrome, and cirrhosis. Furosemide is practically insoluble in water and its oral bioavailability is very poor due to insufficient aqueous solubility at gastrointestinal pH, making solubility the rate-determining step in the gastric absorption of furosemide. Improvement of its solubility and dissolution properties is essential because this closely correlates with its bioavailability.

PREPARATION OF FUROSEMIDE COMPLEXES

Different techniques such as freeze drying, solvent evaporation, and grinding, etc., were used for preparation of complexes. The complexes were prepared in the molar ratio 1:1 and 1:2 for drug:complexing agents. The quantities of furosemide and fulvic/humic acids used for the preparation of complexes in different molar ratio are shown in Table 7.3.

Table 7.3: Quantity of furosemide and complexing agents used for complex preparation

Ratio (drug:complexing agent)	Quantity of furosemide (g)	Quantity of humic acid[a] (g)	Quantity of fulvic acid[b] (g)
1 : 1	330.7	6500	1200
1 : 1	330.7	13000	2400

[a]Average molecular weight, 6500.
[b]Average molecular weight, 1200.

The prepared complexes were characterized by means of differential scanning calorimetry, X-ray diffractometry, and Fourier transform infrared spectroscopy techniques.

DIFFERENTIAL SCANNING CALORIMETRY

DSC thermogram (instrument calibrated by using Indium as a standard having melting point at 165 °C) was recorded using a Perkin-Elmer differential scanning calorimeter equipped with computerized data station.

Furosemide-Fulvic acid complexes prepared by different techniques

Figure 7.29 shows the DSC thermograms for furosemide and fulvic acid in comparison to their complexes prepared in different ratios by different techniques. As shown in the figure, furosemide showed a sharp endotherm at 214 °C while fulvic acid did not exhibit any sharp endotherm, indicating that it does not have any defined melting point. On the other hand, complexes prepared by both solvent evaporation as well as freeze-drying techniques showed the absence of the sharp endotherm corresponding to the melting point of furosemide, indicating complex formation.

Furosemide-humic acid complexes prepared by different techniques

Figure 7.30 shows the DSC thermograms for furosemide and humic acid in comparison to their complexes prepared in different ratios by different techniques. As shown in the figure, furosemide showed a sharp endotherm at 214 °C while humic acid did not exhibit any sharp endotherm, indicating that it does not have any defined melting point. On the other hand, complexes prepared by both solvent evaporation as well as freeze-drying techniques showed the absence of the sharp endotherm corresponding to the melting point of furosemide, indicating complex formation.

POWDER X-RAY DIFFRACTION

X-ray diffraction pattern of furosemide, fulvic acid, and humic acid and their complexes prepared by different techniques. As

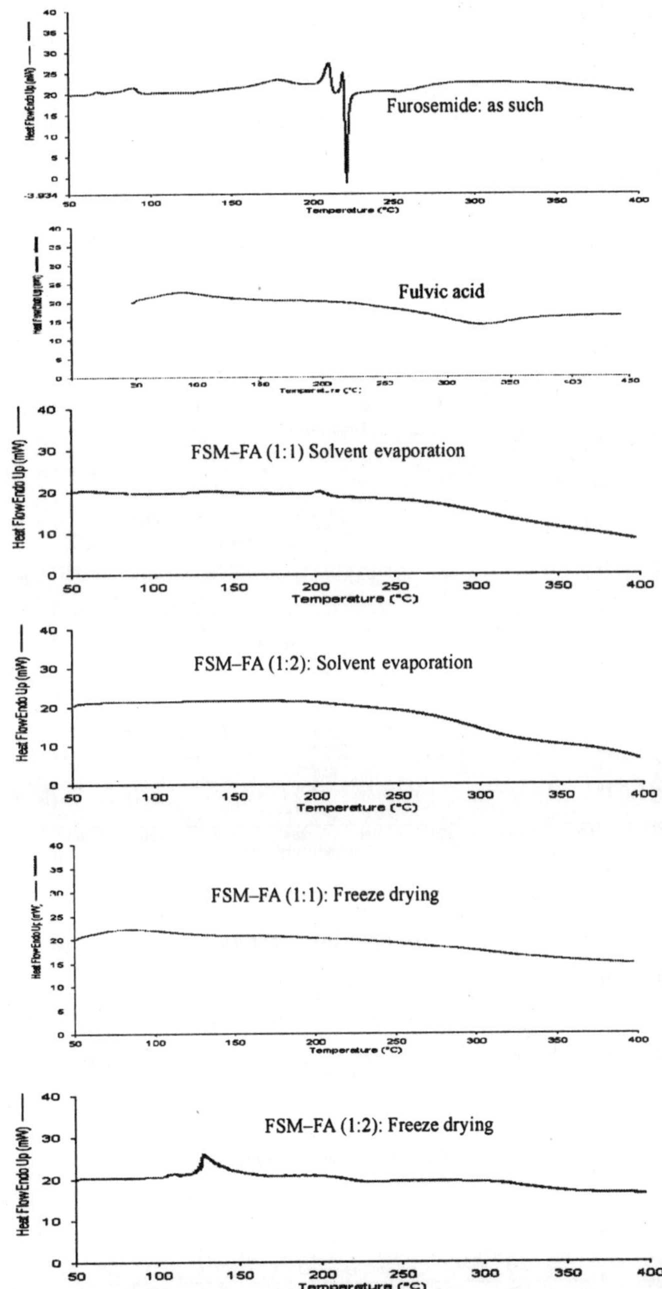

Fig. 7.29: DSC thermogram of furosemide (FSM), fulvic acid (FA), and furosemide-fulvic acid complexes prepared by various techniques

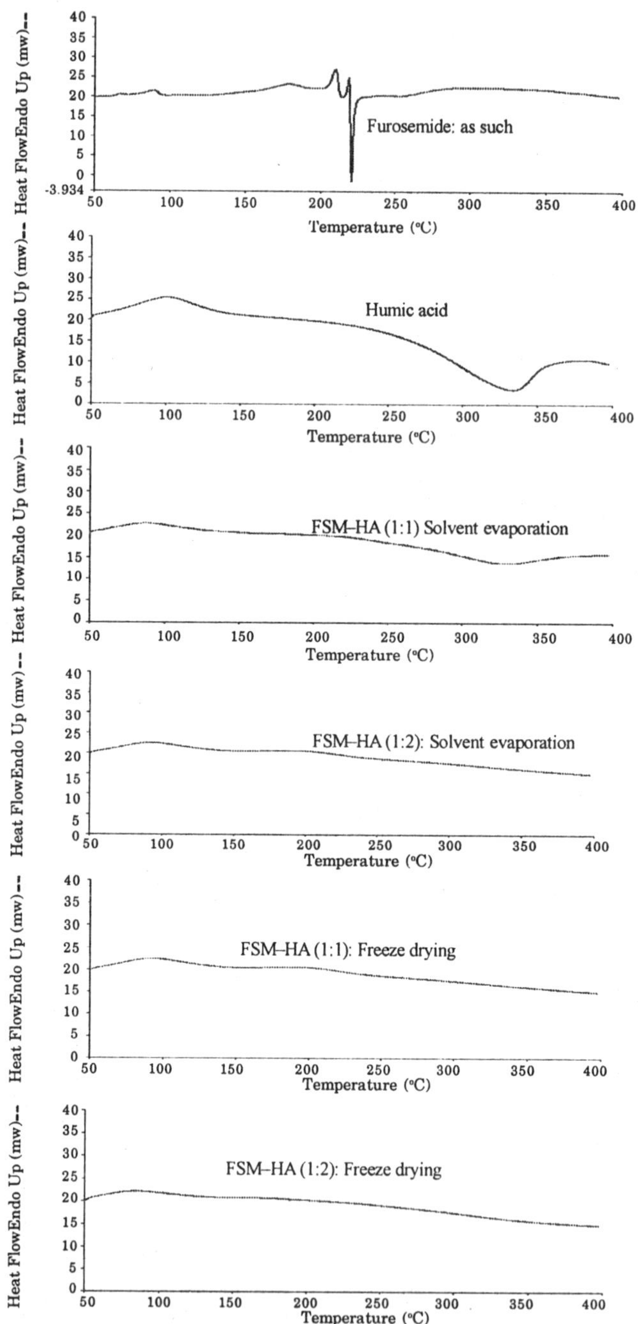

Fig. 7.30: DSC thermogram of furosemide (FSM), humic acid (HA), and furosemide-humic acid complexes prepared by various techniques

shown in the figure, the XRD diffractogram showed a number of sharp peaks characteristic of crystalline nature of furosemide. On the other hand, the XRD diffractogram of complexes prepared with both fulvic acid and humic acid showed the absence of these sharp peaks suggesting complex formation between the drug and the humic substances.

FOURIER TRANSFORM INFRA RED SPECTROSCOPY

FTIR spectra of furosemide-humic acid complex prepared by different techniques

The FTIR spectra of furosemide, fulvic acid, and various complexes of furosemide with fulvic acids in the molar ratio of 1:1 and 1:2 prepared by different techniques. Since FTIR is a highly sensitive method of analysis, all spectra of complexes show some changes from parent spectra, *i.e.*, pure drug and fulvic acid. The FTIR spectra of the furosemide-fulvic acid complex in the molar ratio 1:1 prepared by solvent evaporation showed peaks of furosemide in fingerprint region (1600–400 cm^{-1}) with decreased intensity. It supports that weak interaction of furosemide with fulvic acid. While FTIR of 1:2 molar ratio shows complete peak overlapping of furosemide and fulvic acid at 3900 cm^{-1}, the fingerprint region peaks (1600–400 cm^{-1}) disappear, supporting complex formation.

The FTIR spectra of furosemide-fulvic acid complex prepared by freeze-drying method in the molar ratio 1:1 showed reduced number of furosemide peaks, indicating weak interaction of furosemide and fulvic acid. In the FTIR of furosemide-fulvic acid complex prepared by freeze-drying method in the molar ratio 1:2 showed the complete absence of furosemide peaks. This suggests that complex formation has taken place.

FTIR spectra of furosemide-humic acid complex prepared by different techniques

The FTIR spectra of furosemide, humic acid, and furosemide-humic acid complexes prepared by various techniques in the molar ratio 1:1 and 1:2.

Since FTIR is a highly sensitive method of analysis, all spectra of complexes show some changes from parent spectra, *i.e.*, pure drug and humic acid. The FTIR spectra of the furosemide-humic acid complex in the molar ratio 1:1 and 1:2 prepared by solvent evaporation showed peaks of furosemide in fingerprint region (1600–400 cm^{-1}) with decreased intensity, indicating a weak interaction between furosemide with humic acid.

The FTIR spectra of furosemide-humic acid complex prepared by freeze drying in the molar ratio 1:1 showed peaks of furosemide at 1690 and 1550 cm^{-1}, indicating weak interaction of furosemide and humic acid. In the FTIR spectra of furosemide-humic acid complex prepared by freeze drying in the molar ratio 1:2, the peaks of furosemide were missing indicating complete complex formation.

8

SOLUBILITY AND DISSOLUTION ENHANCING PROPERTIES OF HUMIC SUBSTANCES

It is well recognized that the design and composition of a pharmaceutical dosage form may have an important impact on the bioavailability and hence therapeutic outcome of the drug product. Solubility and permeability are key parameters which determine the fate of an orally administered drug dosage form in the gastrointestinal (GT) tract. Molecules with too low solubility and/or permeability will provide low and variable bioavailability, with the result that a clinically useful product may not be developed.

Drugs are generally given to a patient as a manufactured drug product (finished dosage form) that includes the active drug along with selected inactive ingredients (excipients) that make up the dosage form. The successful transposition of a drug from a solid oral dosage form into the general circulation can be described as four-step process:

(i) Disintegration of the drug product

(ii) Dissolution of the drug in the fluids at the site of absorption

(iii) Absorption of the dissolved drug through the membranes of the GI tract

(iv) Movement of the drug away from the site of absorption into the general circulation.

Any factor which has the ability to affect any of these steps can in turn alter the drug's bioavailability and thereby its therapeutic effect. According to Biopharmaceutics Classification System (BCS), oral bioavailability of Class II drugs (poorly soluble and highly permeable) is often limited by its dissolution rate (Dressman and Reppas, 2000). Therefore, great efforts have been made in the field of pharmaceutical science to increase their bioavailability through enhancement of their solubility and dissolution rates. Some of the approaches that have been used successfully to improve the bioavailability of drugs include micronization, solid dispersion, complexation with cyclodextrins, and use of surfactants.

The authors and coworkers studied the solubility and dissolution enhancing properties of humic substances extracted from shilajit by determining the solubility and dissolution characteristics of the various complexes prepared with humic and fulvic acid. The complexes which showed promising results were further formulated into tablet dosage forms and their *in vitro* dissolution characteristics were determined and compared with uncomplexed drug as well as commercial formulations available in the market containing uncomplexed drug.

I. ITRACONAZOLE

Aqueous Solubility Studies

To determine the effect of complexation on the solubility of itraconazole, the saturation solubility of itraconazole, prepared complexes, and physical mixture were determined in different media (simulated gastric fluid without pepsin, acetate buffer pH 4.0, water and fed state simulated intestinal fluid) at a temperature of 25 ± 2 °C by shake flask method.

An excess quantity of the drug or complex (about 20 mg) was added to 10 mL of media in stoppered glass tubes which were placed in a thermostatically controlled water bath and agitated continuously for 7 days. Preliminary experiments had shown that saturation solubility could be achieved by shaking for 7 days. After 7 days, the solution was centrifuged and the supernatant was filtered through a 0.22 μm membrane filter and analyzed by HPLC method. The concentration of drug in the solution was determined from the calibration curve. The results are shown in Table 8.1.

Table 8.1: Saturation solubility of itraconazole and complexes prepared by different techniques

Sl. No.	Sample	Saturation solubility (µg/mL) in			
		Simulated gastric fluid (pH 1.2)	Acetate buffer (pH 4.0)	Water (pH 6.6)	FeSSIF[a] (pH 5.0)
1.	Itraconazole as such (ITRA)	5.22	0.28	0.06	0.39
2.	ITRA alone (solvent evaporation)	8.54	0.31	0.06	0.42
3.	ITRA alone (freeze drying)	12.87	0.35	0.08	0.45
4.	ITRA alone (spray drying)	30.48	0.41	0.11	0.61
5.	ITRA:FA (1:1) (physical mixture)	4.85	0.27	0.06	0.41
6.	ITRA:FA (1:1) (freeze-dried physical mixture)	5.69	0.29	0.06	0.42
7.	ITRA:FA (1:0.5) (solvent evaporation)	78.36	2.25	0.36	4.36
8.	ITRA:FA (1:1) (solvent evaporation)	143.08	4.42	0.86	8.43
9.	ITRA:FA (1:2) (solvent evaporation)	147.14	4.58	0.88	8.67
10.	ITRA:FA (1:1) (freeze drying)	173.84	4.75	1.15	10.58
11.	ITRA:FA (1:1) (spray drying)	224.88	6.74	1.34	12.32

It is evident from the table that complexation resulted in a significant (more than 20 times) increase in the solubility of the drug, with the maximum increase being observed in case of spray drying. In contrast, there was no significant increase in the solubility in case of physical mixture or when itraconazole alone was freeze dried or spray dried.

Formulation of Itraconazole Tablets Using Complexes

Fast disintegrating tablet formulation incorporating the drug alone or one of the complexes was developed keeping the dispersion time as the critical parameter. A number of diluents including microcrystalline cellulose, lactose, coprocessed microcrystalline cellulose-lactose (Cellactose), starch, dicalcium phosphate, and their combinations in different ratios were tried along with different super-disintegrants like croscarmellose sodium, crospovidone, low-substituted hydroxyl propyl cellulose, etc. A prototype formula was selected based on best dispersion time, physical appearance, and dissolution and was used for further evaluation. Table 8.2 gives the composition of the various prototype formulations developed.

Tablets containing either 100 mg of the uncomplexed drug or containing the physical mixture or complexes equivalent to 100 mg of itraconazole were prepared as per the formula given in Table 8.2. For the preparation of tablets, the ingredients were intimately mixed and compressed on a 16 station rotary tabletting machine.

Release and *In vitro* Equivalence Study

In vitro dissolution studies for the various tablet formulations of itraconazole were carried out in simulated gastric fluid without pepsin at 37 ± 1 °C by the USP paddle method at 100 rpm. This medium and condition have earlier been used by a number of investigators (Jung *et al.*, 1999; Yoo *et al.*, 2000) for determining the dissolution of itraconazole formulations. During the study, samples were withdrawn at 5, 15, 30, 45 and 60 min, filtered through a 0.22 µm membrane filter and were analyzed for the amount of itraconazole dissolved by HPLC method. Fresh aliquots of dissolution medium were added to compensate for the sample withdrawn.

Table 8.2: Composition of prototype tablet formulations of itraconazole

S. No.	Ingredients	Qty/Tab (mg)			
		For Tablet containing uncom-plexed drug	**For Tablet containing ITRA-FA complex 1:0.5**	**For Tablet containing ITRA-FA complex 1:1**	**For Tablet containing ITRA-FA complex 1:2**
1.	Itraconazole	100	–	–	
2.	ITRA-FA physical mixture or complex	–	185	270	440
3.	Cellactose	632	547	462	292
4.	Crospovidone	60	60	60	60
5.	Magnesium stearate	4	4	4	4
6.	Purified talc	4	4	4	4

The dissolution profile of the optimized formulation was compared with that of the innovator's product (SPORANOX capsule 100 mg, M/s Janssen Pharmaceutica, USA) using Wilcoxon signed rank test as well as using the similarity factor (f2) and the difference factor (f1) (Moore and Flanner, 1996).

Itraconazole alone

Table 8.3 and Figure 8.1 show the results of the dissolution studies in simulated gastric fluid of tablets prepared using itraconazole alone as such in comparison with itraconazole alone that has been dissolved in glacial acetic acid-water and has been dried by solvent evaporation in a rotary evaporator (solvent evaporated), freeze drying, and spray drying. As seen from the results, tablets

Table 8.3: Comparative dissolution profile in simulated gastric fluid of tablets containing itraconazole alone: as such, solvent evaporated, freeze dried, and spray dried

Time (min)	% Drug dissolved (±SD) n=3			
	Itraconazole as such	**Itraconazole solvent evaporated**	**Itraconazole freeze dried**	**Itraconazole spray dried**
0	0.0	0.0	0.0	0.0
5	0.8 (0.4)	1.3 (0.5)	2.1 (0.5)	4.7 (1.4)
15	2.9 (0.5)	4.7 (1.5)	5.3 (1.1)	12.4 (1.8)
30	5.4 (0.9)	8.0 (1.5)	11.3 (2.2)	19.4 (2.2)
45	6.7 (0.8)	9.4 (1.4)	13.7 (1.1)	24.6 (2.1)
60	6.8 (0.7)	11.3 (1.0)	16.3 (1.3)	28.7 (2.8)

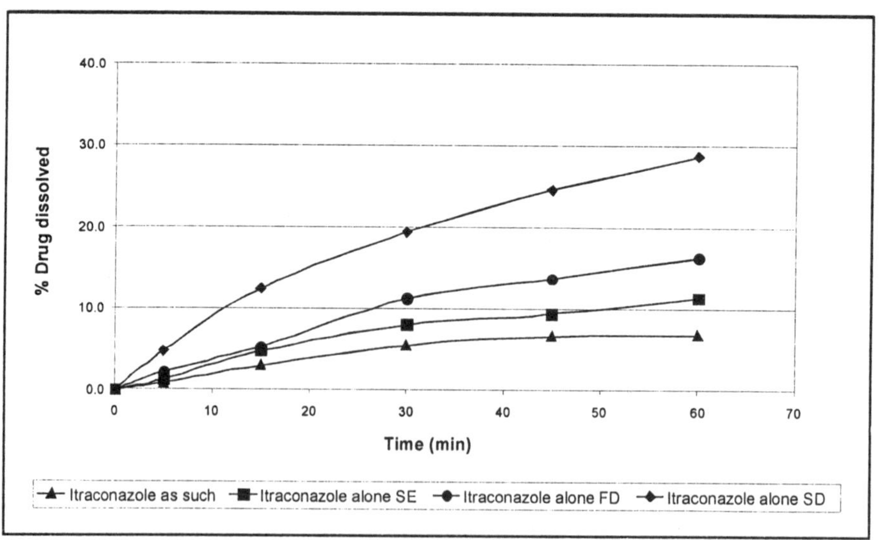

Fig. 8.1: Comparative dissolution profile of tablets containing itraconazole alone: as such, solvent evaporated, freeze dried, and spray dried

prepared using itraconazole as such demonstrate a very poor dissolution, mainly because of its crystalline nature. Although the dissolution of the tablets prepared using itraconazole that had been solvent evaporated, freeze dried, or spray dried was somewhat better than that of itraconazole as such, it was still far less than what was desired. Improvement in the dissolution could be attributed to the conversion of itraconazole from crystalline to the amorphous state due to spray drying, freeze drying, or solvent evaporation in a rotary evaporator.

Itraconazole-fulvic acid complex prepared by physical mixing

Table 8.4 and Figure 8.2 show the results of the dissolution studies in simulated gastric fluid of tablets prepared using itraconazole alone as such in comparison with itraconazole-fulvic acid complexes prepared by physical mixing or freeze drying the physical mixture. As seen from the results, tablets containing physical mixture or freeze-dried physical mixture did not show any improvement in the dissolution as compared to itraconazole alone.

Table 8.4: Comparative dissolution profile in simulated gastric fluid of tablets containing itraconazole alone and itraconazole-fulvic acid complexes by physical mixing

Time (min)	**% Drug dissolved (±SD) n=3**		
	Itraconazole as such	**ITRA:FA (1:1) physical mixture**	**ITRA:FA (1:1) Freeze-dried physical mixture**
0	0.0	0.0	0.0
5	0.8 (0.4)	0.6 (0.5)	1.3 (0.5)
15	2.9 (0.5)	3.7 (1.2)	3.0 (1.2)
30	5.4 (0.9)	6.5 (1.1)	7.0 (0.9)
45	6.7 (0.8)	7.7 (1.0)	8.7 (0.7)
60	6.8 (0.7)	8.0 (0.8)	9.0 (0.6)

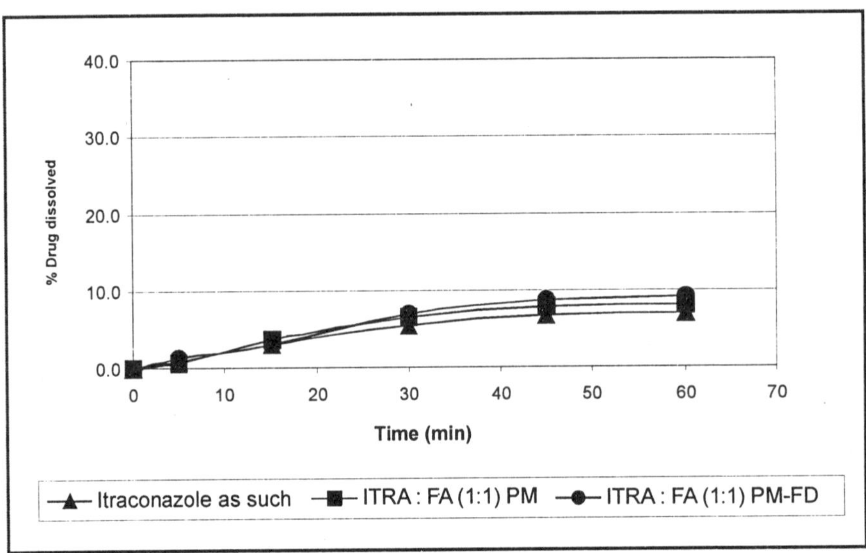

Fig. 8.2: Comparative dissolution profile of tablets containing itraconazole alone and itraconazole-fulvic acid complexes by physical mixing

Itraconazole-fulvic acid complexes prepared by solvent evaporation technique

Table 8.5 and Figure 8.3 show the results of the dissolution studies in simulated gastric fluid of tablets prepared using itraconazole alone as such in comparison with itraconazole-fulvic acid complexes prepared in different ratios by the solvent evaporation techniques. As seen from the results, tablets containing itraconazole-fulvic acid complex in all the ratios showed a better dissolution profile as compared to tablets containing uncomplexed drug. The improvement in 1:1 ratio was significantly more than 1:0.5 ratio. However, there was no significant improvement on increasing the content of fulvic acid from 1:1 to 1:2 ratio. Hence, 1:1 ratio was selected as the optimized ratio in which complexes with other techniques were prepared.

Itraconazole-fulvic acid (1:1) complexes prepared by different techniques

Table 8.6 and Figure 8.4 show the results of the dissolution studies in simulated gastric fluid of tablets prepared using itraconazole

Table 8.5: Comparative dissolution profile of tablets containing Itraconazole alone and complexes prepared in different ratios by solvent evaporation technique

Time (min)	% Drug dissolved (±SD) n=3			
	Itraconazole as such	ITRA:FA (1:0.5) solvent evaporation	ITRA:FA (1:1) solvent evaporation	ITRA:FA (1:2) solvent evaporation
0	0.0	0.0	0.0	0.0
5	0.8 (0.4)	18.4 (2.8)	58.7 (5.7)	53.9 (3.5)
15	2.9 (0.5)	36.9 (4.0)	79.3 (4.8)	74.7 (3.7)
30	5.4 (0.9)	47.7 (3.1)	88.5 (3.1)	87.4 (2.9)
45	6.7 (0.8)	56.4 (4.6)	94.6 (3.4)	95.5 (2.0)
60	6.8 (0.7)	62.1 (3.9)	97.6 (1.1)	98.2 (1.3)

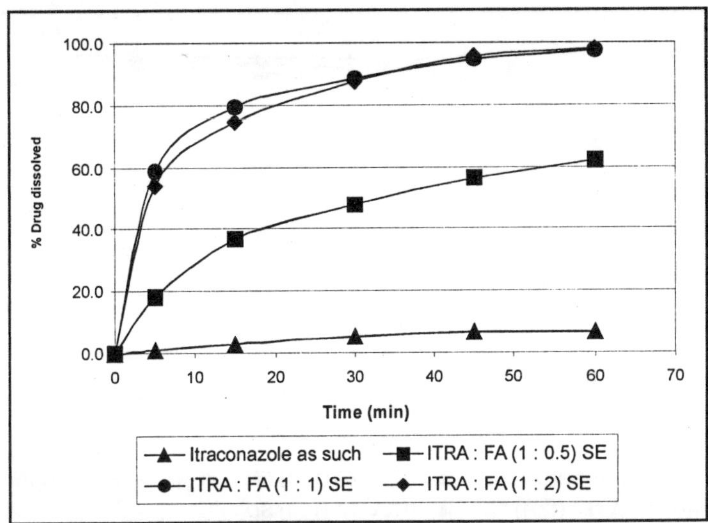

Fig. 8.3: Comparative dissolution profile of tablets containing itraconazole alone and itraconazole-fulvic acid complexes in different ratios prepared by the solvent evaporation technique

Table 8.6: Comparative dissolution profile of tablets containing uncomplexed drug and complexes prepared by different techniques

Time (min)	% Drug dissolved (±SD) n=3				
	Itraconazole as such	**ITRA:FA (1:1) physical mixture**	**ITRA:FA (1:1) solvent evaporation**	**ITRA:FA (1:1) Freeze drying**	**ITRA:FA (1:1) Spray drying**
0	0.0	0.0	0.0	0.0	0.0
5	0.8 (0.4)	0.6 (0.5)	58.7 (5.7)	72.9 (4.3)	76.5 (5.6)
15	2.9 (0.5)	3.7 (1.2)	79.3 (4.8)	90.5 (1.8)	95.9 (2.7)
30	5.4 (0.9)	6.5 (1.1)	88.5 (3.1)	94.6 (2.8)	98.5 (0.9)
45	6.7 (0.8)	7.7 (1.0)	94.6 (3.4)	97.0 (1.9)	99.2 (0.4)
60	6.8 (0.7)	8.0 (0.8)	97.6 (1.1)	97.9 (1.1)	99.8 (0.2)

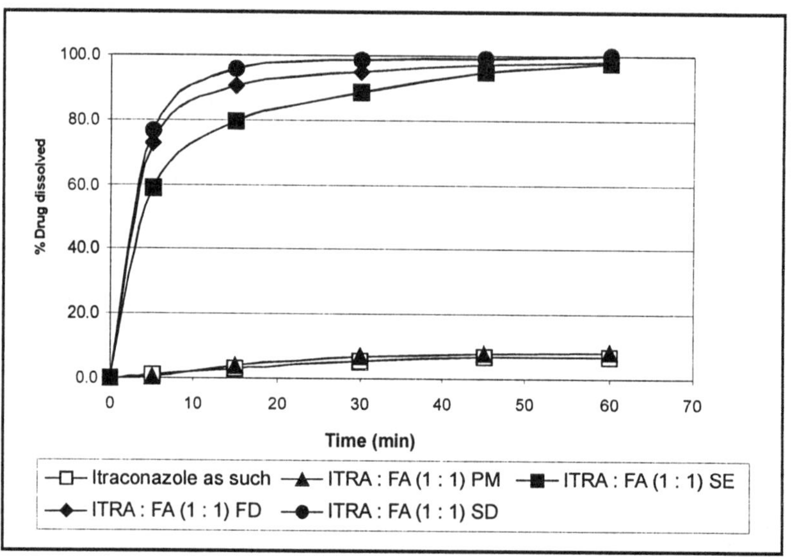

Fig. 8.4: Comparative dissolution profile of tablets containing uncomplexed drug and 1:1 complex prepared by different techniques

alone as such in comparison with itraconazole-fulvic acid complex prepared in 1:1 ratio by various techniques. As seen from the results, the dissolution of tablets containing complexed drug was significantly improved in comparison to that containing uncomplexed drug or a physical mixture of itraconazole and fulvic acid. Amongst the different techniques, the rate and extent of dissolution was the best with 1:1 complex prepared by the spray-drying technique. This combined with the results of solubility studies and results of the characterization studies guided us to select the tablet containing 1:1 complex by the spray-drying technique as the optimized formulation.

Comparative dissolution profiles of optimized itraconazole formulation and innovator product

The dissolution profile of the optimized itraconazole formulation was also compared with the innovator's product of itraconazole (SPORANOX capsule 100 mg, M/s Janssen Pharmaceutica, USA).

Table 8.7 and Figure 8.5 show the comparative dissolution profiles in simulated gastric fluid of tablets prepared using

Table 8.7: Comparative dissolution profile in simulated gastric fluid of optimized itraconazole formulation in comparison to tablets containing uncomplexed drug and innovator formulation

Time (min)	% Drug dissolved (±SD) n=3		
	Itraconazole as such	ITRA:FA (1:1) Spray drying	Sporanox capsules 100 mg
0	0.0	0.0	0.0
5	0.8 (0.4)	76.5 (5.6)	6.1 (4.0)
15	2.9 (0.5)	95.9 (2.7)	34.6 (3.7)
30	5.4 (0.9)	98.5 (0.9)	65.5 (3.6)
45	6.7 (0.8)	99.2 (0.4)	79.4 (3.2)
60	6.8 (0.7)	99.8 (0.2)	86.0 (2.9)

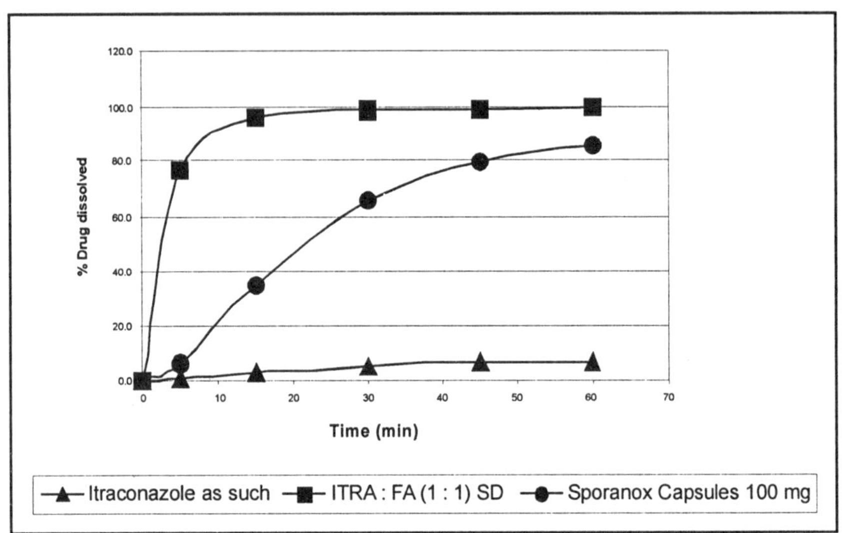

Fig. 8.5: Comparative dissolution profile in simulated gastric fluid of optimized formulation in comparison to tablets containing uncomplexed drug and innovator formulation

Table 8.8: Wilcoxon signed rank test for comparison of % drug release of optimized formulation and innovator formulation in simulated gastric fluid

Time (min)	% Drug release for optimized formulation	% Drug release for sporanox capsule	Difference	Rank of difference (with appropriate sign)	Rank with less-frequent sign
5	76.5	6.1	70.4	+1	NIL[a]
15	95.9	34.6	61.3	+2	
30	98.5	65.5	33.0	+3	
45	99.2	79.4	19.8	+4	
60	99.8	86.0	13.8	+5	

[a]As the sum of the less-frequent sign is 0, the difference is significant at 5% level.

itraconazole alone as such in comparison with the optimized formulation as well as the innovator product. As seen from the results, the dissolution of sporanox capsules was significantly better than the uncomplexed drug while the rate and extent of dissolution of the optimized formulation was still better than both uncomplexed drug as well as the innovator product. The percent drug release values for the optimized tablet was found to be significantly higher (Wilcoxon signed rank test) (Table 8.8) than the innovator product. Comparison of the two profiles by the similarity test (f2 = 17.05) as well as the difference factor (f1 = 53.48) also showed that the release profile of developed formulation was different from the innovator's product.

II. KETOCONAZOLE

Aqueous Solubility Studies

The saturation solubility of ketoconazole, prepared complexes with humic and fulvic acid and the corresponding physical mixtures were determined in different media-simulated gastric fluid without pepsin, acetate buffer pH 4.0, 0.1 M phosphate buffer of pH 5 and 6 and in fed state simulated intestinal fluid (pH 5.0), at a temperature of 25 ± 2 °C by shake flask method (Karmarkar, 2007). The complexes were kept for shaking in the dissolution media for 7 days and saturation solubility was determined by HPLC. The results are shown in Table 8.9.

As is evident from the table, the complexation resulted in a significant increase in the solubility of the drug, with maximum increase observed in case of spray-dried complex. In contrast, there was no significant increase in solubility in case of physical mixture or when ketoconazole alone was dissolved in glacial acetic acid-water mixture and dried by different methods.

Formulation of Ketoconazole Tablets Using Complexes

Keeping the dispersion time as the parameter, fast disintegrating tablet formulation of drug alone or 1:0.5 and 1:1 ketoconazole-fulvic acid complex was developed. A number of diluents such as microcrystalline cellulose, lactose, co–processed microcrystalline cellulose-lactose (cellactose), starch, and their combination in different molar ratios were tried along with different super-

Table 8.9: Saturation solubility of ketoconazole and complexes prepared by different techniques

Sl. No.	Sample	Water (µg/mL)	Simulated gastric fluid (pH 1.2) (µg/mL)	Acetate buffer (pH 4.0) (µg/mL)	0.1 M phosphate buffer (pH 5.0) (µg/mL)	0.1 M phosphate buffer (pH 6.0) (µg/mL)	Fed state simulated intestinal fluid (pH 5.0) (µg/mL)
1	Ketoconazole	0.112 ± 00.03	1210 ± 30.00	86.20 ± 5.20	77.01 ± 3.20	22.01 ± 1.20	79.0 ± 2.16
2	Ketoconazole evaporation)	0.142 ± 00.05	1250 ± 205.00	87.40 ± 3.40	78.14 ± 5.10	24.32 ± 1.68	82.61 ± 3.13
3	Ketocanozole alone (spray dried)	0.157 ± 00.06	1260 ± 31.00	86.50 ± 4.10	79.72 ± 3.90	28.41 ± 3.21	83.19 ± 2.52
4	Ketoconazole: fulvic acid (1:0.5) solvent evaporated complex	908.000 ± 10.32	1156 ± 32.40	180.31 ± 8.14	154.51 ± 7.31	64.31 ± 4.31	92.68 ± 2.15
5	Ketoconazole: fulvic acid (1:0.5) spray-dried complex	915.310 ± 05.45	1352 ± 22.40	198.39 ± 6.90	175.48 ± 8.30	95.71 ± 4.23	111.21 ± 5.20
6	Ketoconazole: fulvic acid (1:1) solvent evaporated complex	914.000 ± 14.31	1452 ± 12.14	212.32 ± 5.14	178.56 ± 4.89	89.41 ± 4.56	110.31 ± 3.56
7	Ketoconazole: fulvic acid (1:1) spray-dried complex	918.000 ± 10.32	1531 ± 21.23	280.31 ± 6.89	245.39 ± 6.14	122.68 ± 8.31	135.63 ± 4.56

disintegrants such as croscarmellose sodium, crospovidone, low-substituted hydroxy propyl cellulose, etc. A prototype formula was selected based on best dispersion time, dissolution, and physical appearance and was used for further evaluation. Table 8.10 gives the composition of the prototype formulation developed.

Tablets containing either 200 mg of the uncomplexed drug or containing the physical mixture or complexes equivalent to 200 mg of ketoconazole were prepared as per the formula given in Table 8.10. For the preparation of tablets, the ingredients were intimately mixed and compressed on a 16 station rotary tabletting machine.

Release and *In vitro* Equivalence Study

Dissolution studies were carried out at 37 ± 1 °C by the USP paddle method in 0.1 M phosphate buffer of pH 5 at 100 rpm on

Table 8.10: Composition of prototype tablet formulations of ketoconazole

S. no.	Ingredients	Quantity per Tab (mg)		
		Uncomplexed drug (mg)	Ketoconazole-fulvic acid (1:1) complex (mg)	Ketoconazole-fulvic acid (1:0.5) complex (mg)
1	Ketoconazole	200	–	–
2	Ketoconazole-fulvic acid physical mixture	–	652	426
3	Cellactose	928	476	702
4	Croscarmellose sodium	60	60	60
5	Magnesium stearate	6	6	6
6	Purified talc	6	6	6

the prepared formulation containing uncomplexed drug, ketoconazole fulvic acid physical mixture, as well as complexes prepared by various techniques. Samples were withdrawn at 5, 15, 30, 45, 60, 120, and 180 minutes, filtered through a 0.22 μm membrane filter and were analyzed for the amount of ketoconazole dissolved by HPLC method. Fresh aliquots of dissolution medium was added as dilution control to compensate for the sample withdrawn.

Ketoconazole alone

The study was conducted to study the effect of process on the dissolution of ketoconazole. Table 8.11 and Figure 8.6 show the result of the dissolution studies in 0.1 M phosphate buffer of pH 5 of tablets prepared using ketoconazole alone, as such in comparison with ketoconazole alone that has been dissolved in glacial acetic acid: water and has been dried by solvent evaporation and spray-drying method.

Table 8.11: Comparative dissolution profile in 0.1M phosphate buffer (pH 5) of tablets containing ketoconazole alone: as such, solvent evaporated, and spray dried

Time (min)	% Drug dissolved (± SD), n = 3		
	Ketoconazole as such	Ketoconazole solvent evaporated	Ketoconazole spray dried
0	0.0	0.0	0.0
5	8.56 (0.3)	9.87 (0.5)	11.35 (0.7)
15	21.54 (0.7)	22.54 (0.3)	22.58 (0.6)
30	28.75 (0.5)	29.57 (0.4)	31.51 (0.8)
45	34.56 (0.6)	35.67 (0.7)	35.68 (0.9)
60	35.89 (0.3)	36.47 (0.5)	36.48 (0.4)
120	35.86 (0.8)	36.21 (0.3)	38.54 (0.7)
180	35.96 (0.6)	37.86 (0.8)	38.94 (0.6)

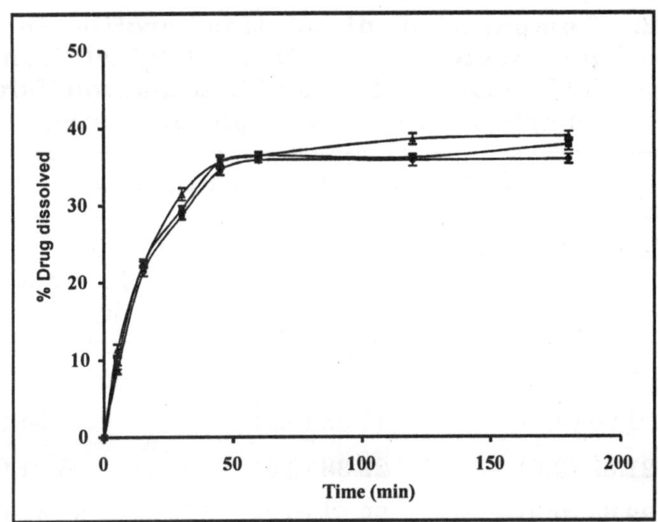

Fig. 8.6: Comparative dissolution profile in 0.1 M phosphate buffer (pH 5) of tablets containing ketoconazole alone: as such (●), solvent evaporated (■), and spray dried (▲)

As seen from the results, tablets prepared using ketoconazole alone or with solvent evaporation and spray-dried ketoconazole showed poor dissolution in dissolution media of pH 5. Also the dissolution profiles in all the tablets were similar which indicates that there is no change in the crystalline nature of ketoconazole by solvent evaporation or with spray drying.

Ketoconazole-fulvic acid complex prepared by physical mixing

Table 8.12 and Figure 8.7 show the results of the dissolution study in 0.1 M phosphate buffer of pH 5 of tablets containing 1:0.5 and 1:1 complex prepared by physical-mixing method. As seen from the results, tablets containing physical mixture did not show any improvement in the dissolution as compared to ketoconazole alone.

Ketoconazole-fulvic acid complex prepared in 1:0.5 and 1:1 molar ratios by solvent evaporation

Table 8.13 and Figure 8.8 shows the results of the dissolution studies in 0.1 M phosphate buffer of pH 5.0 of tablets containing

Table 8.12: Comparative dissolution profile in 0.1 M phosphate buffer (pH 5) of tablets containing ketoconazole alone and ketoconazole-fulvic acid complexes prepared by physical mixing

Time (min)	% Drug dissolved (± SD), n = 3		
	Ketoconazole as such	Ketoconazole-fulvic acid (1:0.5) physical mixture	Ketoconazole-fulvic acid (1:1) physical mixture
0	0.0	0.0	0.0
5	8.96 (0.5)	10.36 (0.3)	12.54 (0.6)
15	21.89 (0.6)	22.38 (0.8)	23.81 (0.4)
30	29.63 (0.3)	29.51 (0.4)	32.45 (0.5)
45	34.36 (0.7)	35.46 (0.6)	36.49 (0.5)
60	36.12 (0.9)	36.52 (0.5)	37.59 (0.6)
120	36.67 (0.3)	36.58 (0.8)	38.47 (0.7)
180	36.98 (0.7)	38.45 (0.9)	39.09 (0.4)

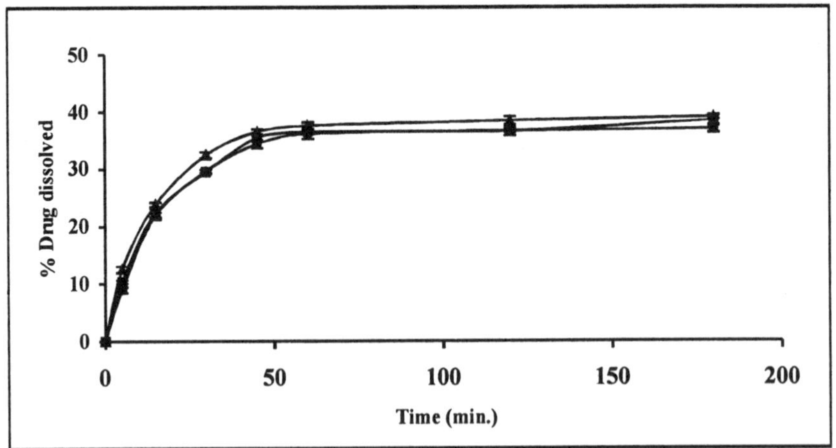

Fig. 8.7: Comparative dissolution profile in 0.1 M phosphate buffer (pH 5.0) of tablet containing ketoconazole alone (•) and ketoconazole-fulvic acid complexes prepared by physical mixing in a molar ratio of 1:0.5 (▪) and 1:1 (▲)

Table 8.13: Comparative dissolution profile of tablets containing ketoconazole alone and complexes prepared in a molar ratio of 1: 0.5 and 1:1 in 0.1 M phosphate buffer of pH 5.0

Time (min)	% Drug dissolved (± SD), n = 3		
	Ketoconazole as such	Ketoconazole-fulvic acid (1:0.5) solvent evaporated complex	Ketoconazole-fulvic acid (1:1) solvent evaporated complex
0	0.0	0.0	0.0
5	11.52 (0.5)	21.67 (2.4)	38.49 (3.5)
15	24.56 (0.8)	43.84 (3.2)	58.69 (2.9)
30	31.56 (0.6)	54.94 (1.8)	70.59 (3.7)
45	35.49 (0.3)	62.81 (3.6)	75.69 (2.3)
60	36.02 (0.2)	66.89 (4.1)	77.15 (2.8)
120	36.45 (0.7)	69.84 (3.9)	77.59 (3.5)
180	36.89 (0.5)	71.56 (3.6)	78.59 (3.7)

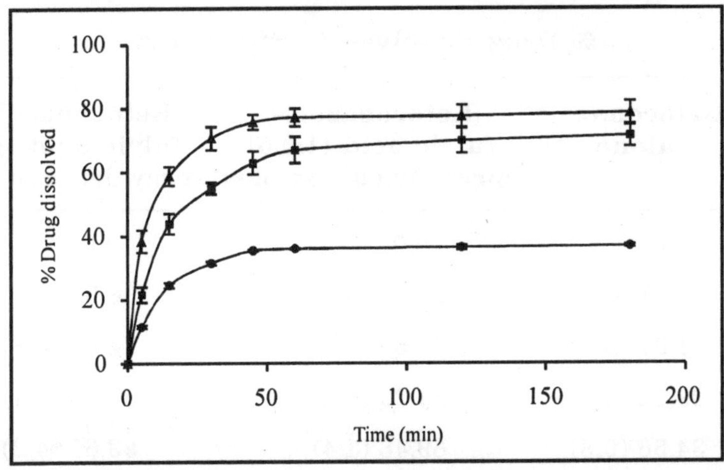

Fig. 8.8: Comparative dissolution profile in 0.1 M phosphate buffer (pH 5.0) of tablet containing ketoconazole alone (•) and ketoconazole-fulvic acid complexes prepared by solvent evaporation in a molar ratio of 1:0.5 (▪) and 1:1 (▲)

1:0.5 and 1:1 complex prepared by solvent evaporation. As seen from the results, tablet containing ketoconazole-fulvic acid complexes showed better dissolution as compared to tablet containing ketoconazole only. Also the dissolution was significantly improved in 1:1 ratio as compared to 1:0.5 ratio.

Ketoconazole-fulvic acid complex prepared in 1:0.5 and 1:1 molar ratios by spray drying

Table 8.14 and Figure 8.9 show the results of the dissolution studies in 0.1 M phosphate buffer of pH 5 of tablets containing 1:0.5 and 1:1 complex prepared by spray drying. As seen from the results, tablets containing ketoconazole-fulvic acid complex showed better dissolution as compared to tablet containing ketoconazole only. Also, the dissolution is almost complete in case of tablet containing 1:1 complex as compared to tablet having 1:0.5 ratio complex in phosphate buffer of pH 5.0.

Table 8.14: Comparative dissolution profile of tablet containing ketoconazole alone and ketoconazole-fulvic acid spray-dried complexes prepared in a molar ratio of 1:1 and 1:0.5 in 0.1 M phosphate buffer of pH 5.0

Time (min)	% Drug dissolved (± SD), n = 3		
	Ketoconazole alone	Ketoconazole-fulvic acid (1:0.5) spray dried complex	Ketoconazole-fulvic acid (1:1) spray dried complex
0	0.0	0.0	0.0
5	9.27 (0.8)	26.51 (0.4)	38.94 (1.6)
15	21.31 (0.5)	49.59 (4.1)	74.37 (3.8)
30	29.12 (0.9)	61.29 (3.8)	87.19 (3.1)
45	34.56 (0.8)	69.48 (1.4)	93.65 (4.2)
60	36.08 (0.5)	75.69 (2.8)	95.89 (1.8)
120	36.58 (0.6)	81.69 (2.7)	96.75 (2.1)
180	36.89 (0.9)	82.56 (1.5)	98.59 (0.8)

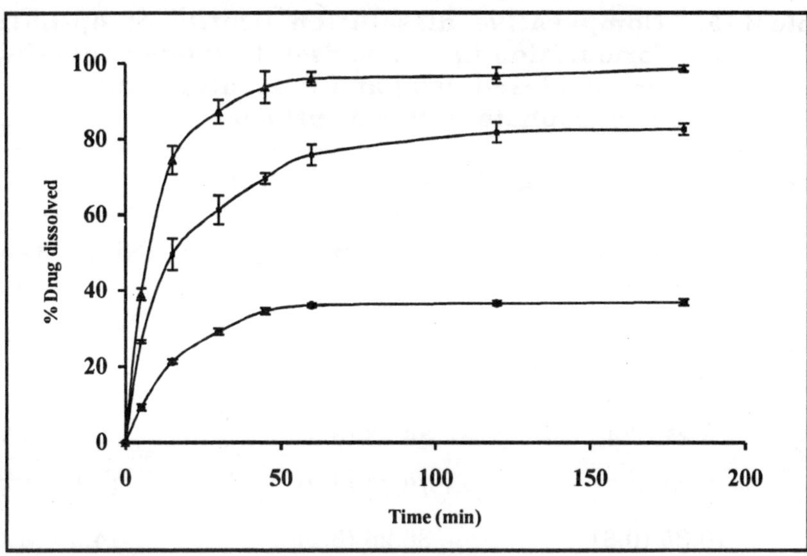

Fig. 8.9: Comparative dissolution profile in 0.1M phosphate buffer (pH 5.0) of tablet containing ketoconazole alone (•) and ketoconazole-fulvic acid complexes prepared by spray drying in a molar ratio of 1:0.5 (▪) and 1:1 (▲)

As seen from the result, the dissolution of tablets containing complexed drug was significantly improved in comparison to that containing ketoconazole alone or a physical mixture of ketoconazole and fulvic acid. Amongst the different techniques, the rate and extent of dissolution was the best in tablet containing 1:1 spray-dried complex. The combined results of characterization and dissolution studies guided us to select the tablet containing 1:1 complex prepared by the spray-drying technique as the optimized formulation.

Comparative dissolution profile of optimized formulation and innovator product

The dissolution profile of the optimized formulation was compared with the innovators product of ketoconazole (Nizrol tablet 200 mg, M/s Jenssen Pharmaceutica, India).

Table 8.15 and Figure 8.10 show the dissoultion profile of tablets prepared using ketoconazole as such in comparison with the optimized formulation, as well as the innovater product in 0.1

Table 8.15: Comparative dissolution profile of optimized formulation in comparison to tablets containing uncomplexed drug and innovator product in 0.1 M phosphate buffer of pH 5.0

Time (min)	% Drug dissolved (± SD), n = 3		
	Ketoconazole as such	**Ketoconazole-fulvic acid (1:1) spray dried complex**	**Nizrol tablet 200 mg**
0	0.0	0.0	0.0
5	3.21 (0.4)	34.48 (3.8)	6.98 (0.7)
15	7.24 (0.7)	72.25 (4.2)	11.58 (1.4)
30	10.35 (0.6)	88.26 (3.1)	14.56 (2.1)
45	12.66 (0.7)	94.75 (2.3)	16.84 (0.9)
60	13.89 (0.5)	96.21 (1.8)	17.59 (0.7)
120	14.31 (0.2)	96.75 (1.4)	17.95 (0.8)
180	14.81 (0.8)	98.12 (0.9)	18.05 (0.9)

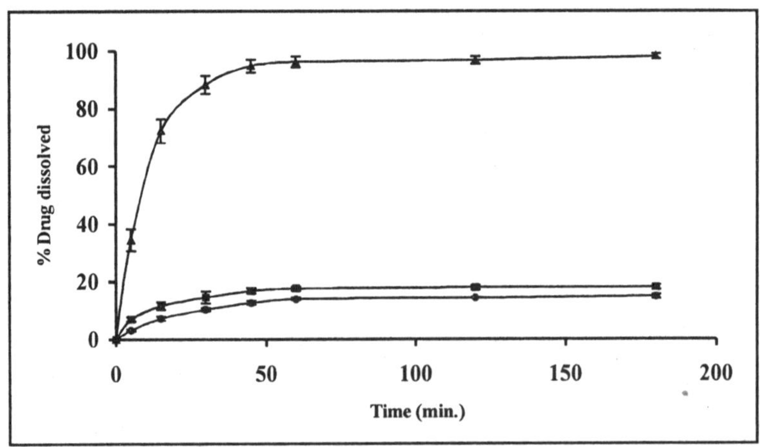

Fig. 8.10: Comparative dissolution profile of optimized formulation (▲) in comparison to tablet containing uncomplexed drug (●) and innovator product (■) in 0.1 M phosphate buffer of pH 5.0

M phosphate buffer of pH 5.0. As seen from the results, the dissolution of Nizrol tablet was almost similar as that of ketoconazole alone tablet. The dissolution efficiency of the optimized formulation was signifcantly higher ($F(2,18) = 60.94$ and 27.91, $a < 0.05$) as compared to innovator product and ketoconazole alone tablet.

III. FUROSEMIDE

Aqueous Solubility Studies

The saturation solubility of furosemide, prepared complexes with fulvic acid and the corresponding physical mixtures were determined in distilled water at a temperature of 25 ± 2 °C by shake flask method (Anwar, 2005). The complexes were kept for shaking in the dissolution media for 5 days and saturation solubility was determined by HPLC. The results are shown in Table 8.16.

Table 8.16: Aqueous solubility determination of furosemide-fulvic acid complexes

Sl. No.	Sample	Solubility of furosemide (μg/mL)	Increase in solubility as compared to drug as such
1	Furosemide as such	32.2	0 times
2	Furosemide: fulvic acid (1:1) physical mixture	157.5	~5 times
3.	Furosemide: fulvic acid (1:2) physical mixture	258.5	~8 times
4.	Furosemide: fulvic acid (1:1) solvent evaporated complex	286.8	~9 times
5	Furosemide: fulvic acid (1:2) solvent evaporated complex	293.0	~9 times
6	Furosemide: fulvic acid (1:1) spray-dried complex	525.3	~16 times
7	Furosemide: fulvic acid (1:2) spray-dried complex	743.3	~23 times

Furosemide is reported to be practically insoluble in water, its saturated solubility in distilled water at room temperature was found to be 32.25 µg/mL. Complex formation of furosemide with fulvic acid greatly enhanced its aqueous solubility with maximum aqueous solubility obtained in case of 1:2 freeze-dried complex.

In vitro release studies of furosemide from complexes

In vitro release studies of pure furosemide (40 mg) and various complexes (equivalent to 40 mg furosemide) were performed using USP Type I (Paddle) tablet dissolution apparatus at 50 rpm in 900 mL phosphate buffer (pH 5.8) at 37.5 ± 1 °C. The samples were withdrawn with the help of syringe fitted with needle and filtered through 0.22 µm membrane filter. Fresh aliquots of the dissolution medium were added to compensate for the quantity of sample withdrawn. The filtered samples were analyzed by HPLC.

The results of the study are shown in Figure 8.11. As can be seen from the figure, the dissolution rate of furosemide alone was found to be very slow due to its extremely poor aqueous solubility. Only 3% release was obtained at 1 hour as compared to a release of 86.9% from the 1:2 freeze-dried complex of furosemide and fulvic acid. No significant increase in the dissolution rate was

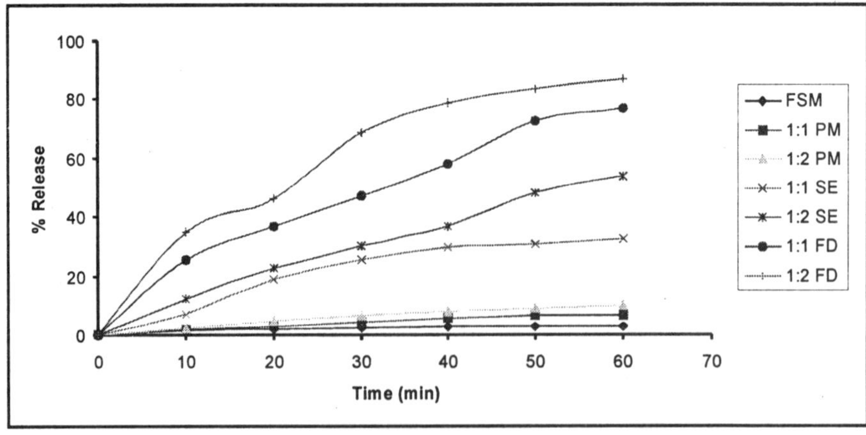

Fig. 8.11: Release profile of furosemide and furosemide-fulvic acid complexes in phosphate buffer pH 5.8

observed with 1:1 and 1:2 physical mixture of furosemide and fulvic acid as compared to drug alone. The study clearly demonstrated that complexation of furosemide with fulvic acid resulted in a significant increase in its dissolution rate.

Formulation of Furosemide Tablets Using Complexes

Based on the release profile of furosemide from various complexes, 1:2 freeze-dried complex of furosemide and fulvic acid was selected for further development based on its better dissolution profile. A conventional release tablet formulation was prepared by direct compression method using the composition shown in Table 8.17.

Table 8.17: Composition of tablet formulation of furosemide

S.no.	Ingredients	Qty/Tab (mg)
1.	Furosemide-fulvic acid complex (1:2)	330
2.	Mannitol	237
3.	Croscarmellose Sodium	30
4.	Magnesium stearate	3

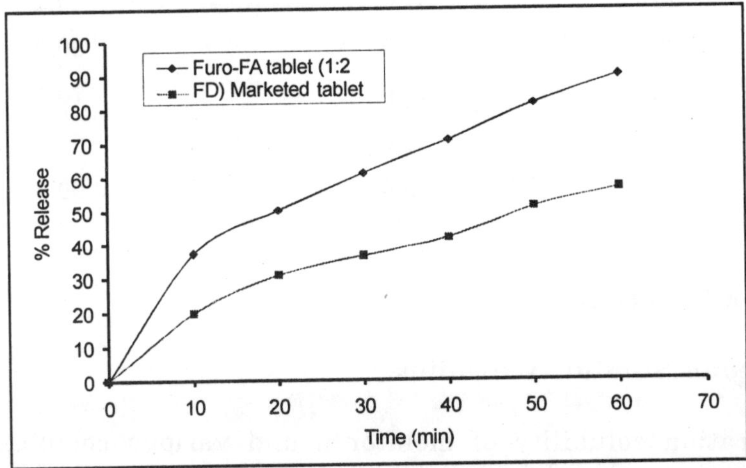

Fig. 8.12: Comparative release profile of developed formulation with marketed preparation of furosemide

Release and *In vitro* Equivalence Study

The *in vitro* dissolution study of the final formulation was carried out in 900 mL of phosphate buffer pH 5.8 at 37.5 ± 1 °C using USP paddle method at 50 rpm and the results were compared with a marketed preparation containing uncomplexed drug (Table 8.18 and Figure 8.12).

As seen from the results, the dissolution of the developed formulation containing furosemide-fulvic acid complex was found to be significantly higher than the marketed formulation containing uncomplexed drug.

Table 8.18: Comparative dissolution profile in phosphate buffer pH 5.8 of optimized furosemide formulation in comparison to marketed formulation containing uncomplexed drug

S.No	Time (min)	% Drug release	
		Marketed formulation containing uncomplexed drug	Tablet containing furosemide-fulvic acid complex (1:2)
1	10	19.9	37.4
2	20	31.0	50.2
3	30	36.5	60.9
4	40	41.9	70.7
5	50	51.0	81.5
6	60	56.5	90.1

IV. MELATONIN

Aqueous Solubility Studies

Saturation solubility of melatonin and various complexes of melatonin with fulvic and humic acid was determined by the shake flask method in distilled water at 25 ± 2 °C (Ahmad, 2006). The complexes were kept for shaking in the dissolution media for

Table 8.19: Aqueous solubility determination of melatonin-fulvic acid complexes

Sl. No.	Sample	Solubility of melatonin (µg/mL)	Increase in solubility as compared to drug as such
1	Melatonin as such	102.7	0 times
2	Melatonin: fulvic acid (1:1) physical mixture	375.2	~3.5 times
3	Melatonin: fulvic acid (1:2) physical mixture	460.9	~4.5 times
4	Melatonin: fulvic acid (1:!) solvent evaporated complex	531.5	~5 times
5	Melatonin: fulvic acid (1:2) solvent evaporated complex	819.8	~8 times
6	Melatonin: fulvic acid (1:!) freeze dried complex	1528.5	~15 times
7	Melatonin: fulvic acid (1:2) freeze dried complex	1801.6	~18 times
8	Melatonin: fulvic acid (1:!) spray dried complex	1331.0	~13 times
9	Melatonin: fulvic acid (1:2) spray dried complex	1555.1	~15.5 times
10	Melatonin: humic acid (1:!) physical mixture	305.1	~3 times
11	Melatonin: humic acid (1:2) physical mixture	418.8	~4 times
12	Melatonin: humic acid (1:!) solvent evaporated complex	475.4	~4.5 times
13	Melatonin: humic acid (1:2) solvent evaporated complex	735.7	~7 times

Table 8.19: Continued ....

Sl. No.	Sample	Solubility of melatonin (µg/mL)	Increase in solubility as compared to drug as such
14	Melatonin: humic acid (1:!) freeze dried complex	1304.1	~13 times
15	Melatonin: humic acid (1:2) freeze dried complex	1521.16	~15 times
16	Melatonin: humic acid (1:!) spray dried complex	1190.74	~11.5 times
17	Melatonin: humic acid (1:2) spray-dried complex	1344.7	~13 times

5 days and saturation solubility was determined by HPLC. The results are shown in Table 8.19.

Melatonin is reported to be very slightly soluble in water and its saturation solubility in distilled water at 25 ± 2 °C was found to be 102.7 µg/mL. Complex formation of melatonin with fulvic acid and humic acid greatly enhanced the aqueous solubility with maximum increase in the aqueous solubility being observed in case of 1:2 freeze-dried complex of melatonin with fulvic acid.

In vitro Release Studies of Melatonin from Complexes

In vitro release studies of pure melatonin (3 mg) and various complexes (equivalent to 3 mg melatonin) were performed using USP Type I (Paddle) tablet dissolution apparatus at 50 rpm in 900 mL phosphate buffer (pH 3) at 37.5 ± 1 °C. The samples were withdrawn with the help of syringe fitted with needle and filtered through 0.22 µm membrane filter. Fresh aliquots of the dissolution medium were added to compensate for the quantity of sample withdrawn. The filtered samples were analyzed by HPLC.

The results of the study are shown in Figures 8.13 and 8.14. As can be seen from the figure, the dissolution rate of melatonin

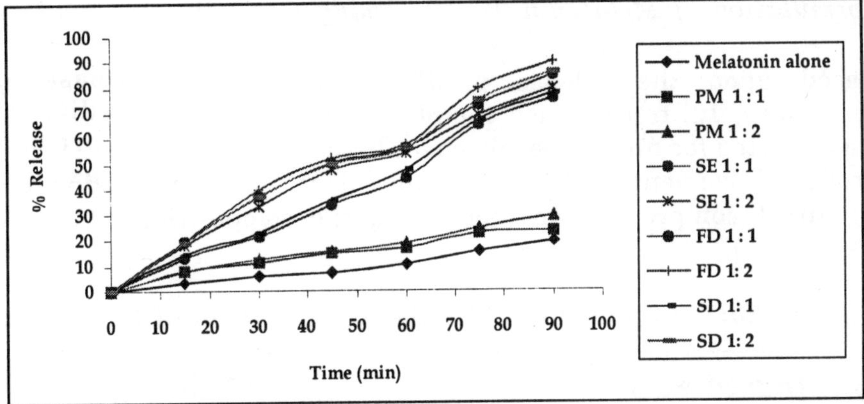

Fig. 8.13: Release profile of melatonin and melatonin-fulvic acid complexes in phosphate buffer pH 3.0

alone was found to be very slow due to its poor aqueous solubility. Only 20% release was obtained in 90 minutes as compared to a release of upto 90% from the 1:2 freeze-dried complex of melatonin and fulvic acid. No significant increase in the dissolution rate was observed with 1:1 and 1:2 physical mixture of melatonin and fulvic acid or humic acid as compared to drug alone. The study clearly demonstrated that complexation of melatonin with fulvic acid or humic acid resulted in a significant increase in its dissolution rate.

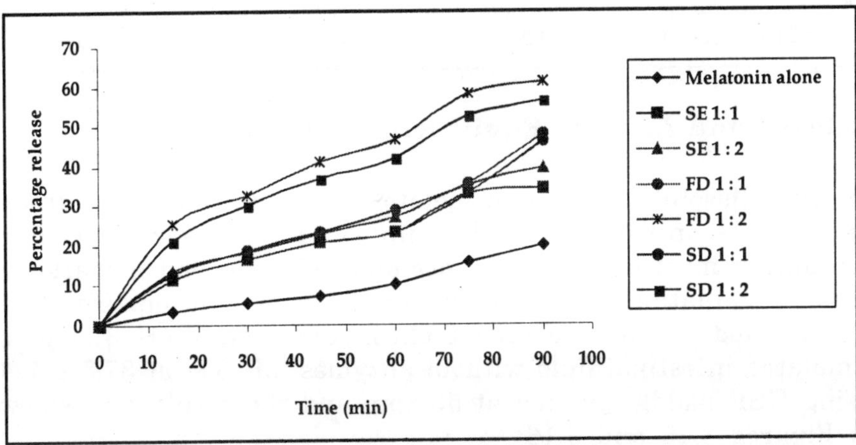

Fig. 8.14: Release profile of melatonin and melatonin-humic acid complexes in phosphate buffer pH (3.0)

Formulation of Melatonin Tablets Using Complexes

Based upon the release profile of melatonin from various complexes, 1:2 freeze dried complex of melatonin and fulvic acid was selected for further development based on its better dissolution profile. A conventional release tablet formulation was prepared by direct compression method using the composition shown in Table 8.20.

Table 8.20: Composition of Tablet formulation of Melatonin

Sl. No.	Ingredients	Qty/Tab (mg)	
		For tablet containing uncomplexed drug	For tablet containing MEL-FA complex 1:2
1.	Melatonin	3.0	–
2.	Melatonin-FA (1:2) complex	–	34.0
3.	Directly compressible microcrystalline cellulose	102.6	71.6
4.	Lactose	60.0	30.0
5.	Crospovidone	12.0	30.0
6.	Magnesium stearate	2.4	3.0

Release and *In vitro* Equivalence Study

In vitro dissolution study of melatonin tablets containing freeze dried melatonin-fulvic acid complex (1:2), prepared tablets containing melatonin in the uncomplexed form and a marketed tablet formulation containing melatonin in the uncomplexed form was carried out in 900 mL of phosphate buffer (pH 3) and in simulated intestinal fluid without enzymes (pH 6.8) at 37.5 ± 1 °C using USP paddle method at 50 rpm and the results are shown in Figures 8.15 and 8.16.

As shown in from the figures, the developed formulation containing freeze-dried melatonin-fulvic acid complex in the molar

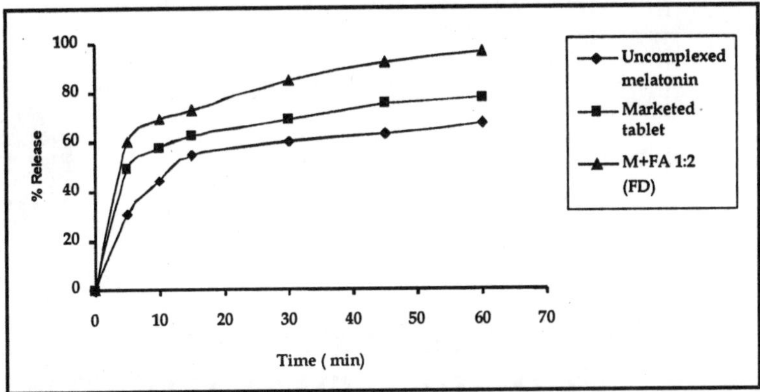

Fig. 8.15: Comparative release profile of developed formulation with marketed preparation of melatonin in phosphate buffer pH 3

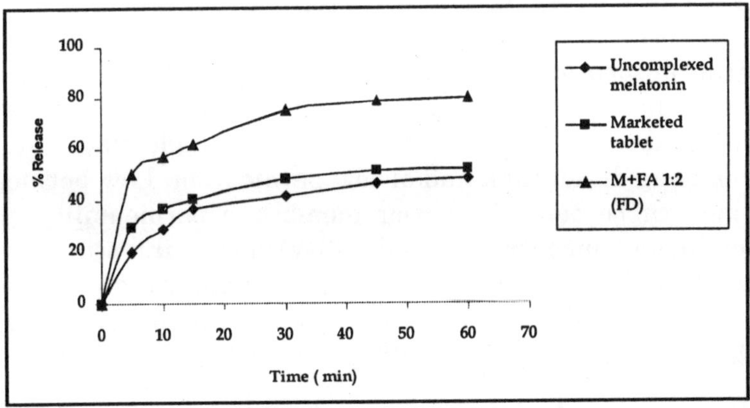

Fig. 8.16: Comparative release profile of developed formulation with marketed preparation of melatonin in simulated intestinal fluid without enzymes pH 6.8

ratio 1:2 showed a release of 96.56% and 80.29% in 1 hour in phosphate buffer (pH 3) and simulated intestinal fluid without enzymes (pH 6.8), respectively. The prepared tablets containing melatonin in the uncomplexed form and marketed tablets containing melatonin in the uncomplexed form showed comparatively slower release than the tablets containing melatonin complexed with fulvic acid. This clearly demonstrates the dissolution-enhancing properties of fulvic acid.

9

PERMEABILITY ENHANCING PROPERTIES OF HUMIC SUBSTANCES

Oral drug absorption can simply be considered as a consecutive process of dissolution and permeation. For a chemically stable drug, poor absorption can be caused by an inadequate rate and extent of drug dissolution and/or low permeation. Low permeation could in turn be caused by poor membrane permeability and/or by a low drug concentration (solubility) in the intestinal lumen. This implies that oral drug absorption or bioavailability can be limited by drug dissolution rate, membrane permeability, or solubility (Shargel and Yu, 1992).

If the drug is hydrophilic with high aqueous solubility, its dissolution in the body is rapid and the rate-determining step in the absorption of such drugs is the rate of permeation across the intestinal membrane. In order to exhibit high oral bioavailability, these drugs must permeate the intestinal mucosa, representing a physical barrier to hydrophilic solutes. The tight junctions between the intestinal mucosal cells restrict the paracellular permeation of solutes depending on their molecular size, charge, and hydrophilicity. Hydrophilic solutes (*e.g.*, peptides) can permeate via the transcellular pathway, if they have affinity for a transporter (*e.g.*, peptide transporter). The permeation of lipophilic solutes by passive diffusion across the intestinal mucosa by the transcellular pathway is dependent on the hydrophobicity and hydrogen bonding potential of the molecule (Calcagno and Siahaan, 2005).

Problems of poor permeability has been sought to be overcome by the use of agents such as surfactants, bile salts, fatty acids and their derivatives, chelators, complexing agents such as cyclodextrins and their derivatives, and polymers like polycarbophil and chitosan. However, the use of these approaches are often limited by several practical factors such as toxicity, irritancy, nonselectivity, poor drug stability, excessive size due to the need for large amount of excipients in relation to the dose, technical manufacturing problems, and the high cost of goods. Thus, there exist a definite need for identifying newer and novel approaches to overcome the problems of poor permeability of drug candidates so as to augment their bioavailability.

It was thought worthwhile to evaluate whether complexation of drugs with humic substances was able to increase the permeability of problematic drugs across the intestinal membrane. Slight modification of the rat everted gut sac method reported by Barthe *et al.* (1998a,b) and Carreno-Gomez *et al.* (2000) was used for carrying out the intestinal permeability studies.

ITRACONAZOLE PERMEABILITY STUDIES

In order to study the effect of complexation on the intestinal permeability of itraconazole, the permeability of itraconazole-fulvic acid complex prepared by spray drying was compared with itraconazole alone as well as with itraconazole-fulvic acid physical mixture by the rat everted gut sac technique. For the study, rat everted intestinal sacs of about 5-cm length were prepared in tissue culture medium, TC199, and filled with about 3 mL of Fed-state simulated intestinal fluid. The sacs were placed in tubes containing 25 ml of Fed-state simulated intestinal fluid in which excess (about 25 mg equivalent of itraconazole) of either itraconazole-alone or itraconazole-fulvic acid (1:1) physical mixture, or itraconazole-fulvic acid (1:1) spray-dried complex had been suspended/dissolved. The tubes were maintained in a shaking water bath at 37 °C, continuously bubbled with oxygen and agitated at a speed of 60 rpm.

Samples were withdrawn from the mucosal side at the start and from the mucosal and serosal side after 1 hour. The samples

Table 9.1: Parameters for permeability study of itraconazole

Sl. No.	Parameter	Condition
1.	Samples	Itraconazole or complex or physical mixture equivalent to 25 mg of itraconazole
2.	Liquid used for washing the lumen of intestine	Normal saline (0.9% sodium chloride at 37 °C)
4.	Liquid used for storing the intestine before mounting	Tissue culture medium TC199 at 37 °C
5.	Mucosal fluid	Fed-state simulated intestinal fluid at 37 °C
6.	Serosal fluid	Fed-state simulated intestinal fluid at 37 °C
7.	Length of the sacs	5 cm
8.	Vol. of mucosal fluid	25 mL
9.	Vol. of serosal fluid	3.0 mL
10.	Oscillations during the study	60/min
11.	Sampling time	1 hr

Table 9.2: Concentration of itraconazole in the mucosal fluid at the start and after 1 hour during permeability studies by everted rat gut sac technique

Sample	Conc. of Itraconazole in the mucosal fluid (μg/mL)	
	Initially	**After 1 hour**
Itraconazole as such (ITRA)	00.32 ± 0.06	00.31 ± 0.09
ITRA:FA (1:1) (physical mixture)	00.34 ± 0.05	00.30 ± 0.07
ITRA:FA (1:1) (spray drying)	11.32 ± 0.43	11.21 ± 0.36

Table 9.3: Conc. of itraconazole in the serosal fluid at the start and after 1 hour during permeability studies by everted rat gut sac technique

Sample	Conc. of Itraconazole in the serosal fluid (µg/mL)	
	Initially	**After 1 hour**
Itraconazole as such (ITRA)	N.D.	0.042 ± 0.005
ITRA:FA (1:1) (physical mixture)	N.D.	0.041 ± 0.007
ITRA:FA (1:1) (spray drying)	N.D.	1.594 ± 0.136

N.D.: Not detected

were centrifuged for 5 min at 4000 rpm, filtered through 0.22 µm membrane filter and analyzed by HPLC method. Table 9.1 summarizes the parameters used for carrying out the permeability studies.

The results of the study are shown in Tables 9.2 and 9.3. As seen from the results, the permeation of itraconazole from itraconazole-fulvic acid complex (1:1) prepared by spray drying was found to be higher than that with itraconazole alone or itraconazole-fulvic acid (1:1) physical mixture. The difference was statistically significant ($P = 0.002$) using one-way repeated measure analysis of variance.

ACYCLOVIR PERMEABILITY STUDIES

In order to determine the effect of complexation on the permeability of BCS Class III drugs, permeability studies were carried out by the rat everted gut sac method using acyclovir as a model drug.

The study was carried out by a method similar to that described for itraconazole except that tissue culture medium TC199 was used as the mucosal as well as serosal liquid. The method is based on the methods reported by Mizuma *et al.* (1999) and, Barthe *et al.* (1998a,b). In order to study the effect of complexation, a 1:1 molar freeze-dried complex of acyclovir and fulvic acid was prepared and compared with acyclovir alone, 1:1

Table 9.4: Parameters for permeability study of acyclovir

Sl. No.	Parameter	Condition
1.	Samples	Acyclovir or complex or physical mixture equivalent to 0.5 mM of acyclovir (112.6 µg/mL)
2.	Intestinal segments	Segment 1: First 15 cm length comprising the duodenum. Segment 2: Next 15 cm length comprising mainly the upper jejunum. Segment 3: Remaining portion of the intestine comprising the lower jejunum and ileum.
3.	Liquid used for washing the lumen of intestine	Normal saline (0.9% sodium chloride at 37 °C)
4.	Liquid used for storing the intestine before mounting	Tissue culture medium TC199 at 37 °C
5.	Mucosal fluid	Tissue culture medium TC199 at 37°C
6.	Serosal fluid	Tissue culture medium TC199 at 37 °C
7.	Length of the sacs	5 cm
8.	Volume of mucosal fluid	25 mL
9.	Volume of serosal fluid	3.0 mL
10.	Oscillations during the study	60/min
11.	Sampling time	1 hr

physical mixture, 1:1 complex of acyclovir and HP-β-cyclodextrin (HP-β-CD) prepared by freeze drying and the innovator product (Zovirax tablet).

Since it has been reported that the acyclovir shows a regional permeability in the different regions of the intestine (Park *et al.*,

1992), the rat intestine was divided into three portions. The first 15 cm length comprising of the duodenum was considered to be the upper intestine followed by the next 15 cm length which was considered to be the middle intestine. The remaining portion of the intestine comprising of the lower jejunum and ileum was considered as the lower intestine. The permeability studies were carried out separately in the three portions. Since it has been reported in the literature that the permeability of acyclovir follows a linear pattern in the concentration range of 5 µM to 5 mM (Fujioka *et al.*, 1991), a single concentration of 0.5 mM (112.6 µg/mL) of acyclovir or complexes equivalent to acyclovir were taken on the mucosal side for the permeability studies. Table 9.4 summarizes the various parameters used for the permeability study of acyclovir.

The results for the study are shown in Table 9.5. As can be seen from the table, acyclovir demonstrated a regional permeability in the different regions of the rat intestine which is in agreement with earlier studies (Park *et al.*, 1992). The permeation of acyclovir, in all the regions of the intestine, from acyclovir-fulvic acid freeze dried complex (1:1) was found to be

Table 9.5: Concentration of acyclovir in the serosal fluid after 1 hour during permeability studies by everted rat gut sac technique

Sample	Conc. (µg/mL) of acyclovir in the serosal fluid after 1 hour		
	Upper intestine	Middle intestine	Lower intestine
Acyclovir as such (ACY)	36.69 ± 3.18	31.58 ± 1.52	23.42 ± 2.85
ACY:FA (1:1) (physical mixture)	38.49 ± 1.87	32.31 ± 2.77	24.63 ± 3.47
ACY:FA (1:1) (freeze drying)	52.97 ± 3.56	48.80 ± 2.53	39.21 ± 3.47
ACY: HP-β-CD (1:1) (freeze drying)	34.65 ± 2.84	30.91 ± 3.04	26.42 ± 2.54
Zovirax tablet	35.74 ± 2.12	29.42 ± 3.98	24.59 ± 3.41

higher than that with acyclovir alone or other complexes of acyclovir as well as the marketed tablet. The difference was statistically significant ($p < 0.05$) using one way repeated measure analysis of variance.

KETOCONAZOLE PERMEABILITY STUDY

In order to study the effect of complexation on the intestinal permeability of ketoconazole, the permeability of ketoconazole-fulvic acid spray-dried complex was compared with ketoconazole alone as well as with ketoconazole-fulvic acid physical mixture by the rat everted gut sac technique.

For the study, rat everted intestinal sac of about 5 cm were prepared and filled with about 3.0 mL of tissue culture medium TC199. The sacs were placed in tubes containing 25 mL of TC199

Table 9.6: Parameters for permeability study of ketoconazole

Sl. No	Parameter	Condition
1	Sample	Ketoconazole or complex or physical mixture equivalent to 30 mg of ketoconazole
2	Liquid used for washing the lumen of intestine	Normal saline (0.9 % sodium chloride at 37 °C).
3	Liquid used for storing the intestine before mounting	Tissue culture medium TC199 at 37 °C.
4	Mucosal fluid	TC199 medium
5	Serosal fluid	TC199 medium
6	Length of the sac	5 cm
7	Volume of the mucosal fluid	25 mL
8	Volume of the serosal fluid	3.0 mL
9	Oscillation during the study	50/min
10	Sampling time	2 h

medium in which excess (about 30 mg equivalent of ketoconazole) of either ketoconazole alone or ketoconazole-fulvic acid (1:1) physical mixture or ketoconazole-fulvic acid (1:1) spray-dried complex had been suspended/dissolved. The tubes were maintained in a shaking water bath at 37 °C at a speed of 50 rpm.

Samples were withdrawn from the mucosal and serosal side at the start and after 2 hours. The samples were centrifuged for

Table 9.7: Concentration of ketoconazole in mucosal side at the start and after 2 hours during permeability study by everted gut sac technique

Sl. No.	Sample	Conc. of ketoconazole in the mucosal fluid (µg/mL)	
		Initial	After 2 hours
1	Ketoconazole as such	0.48 ± 0.04	0.86 ± 0.02
2	Ketoconazole-fulvic acid (1:1) physical mixture	0.49 ± 0.1	0.98 ± 0.05
3	Ketoconazole-fulvic acid (1:1) spray-dried complex	0.49 ± 0.15	8.79 ± 2.1

Table 9.8: Concentration of ketoconazole in the serosal side at the start and after 2 hours during permeability study by everted gut sac technique.

Sl. No.	Samples	Conc. of ketoconazole in serosal fluid (µg/ml)	
		Initial	After 2 hours
1	Ketoconazole as such	N.D.	0.42 ± 0.02
2	Ketoconazole-fulvic acid (1:1) physical mixture	N.D.	0.58 ± 0.10
3	Ketoconazole-fulvic acid (1:1) spray dried complex	N.D.	5.02 ± 0.8

N.D.: Not detected

5 minutes at 2000 rpm, filtered through 0.22 µm membrane filter, and analyzed by HPLC method. Table 9.6 summarized the parameter used for carrying out the permeability studies.

The results for the study are shown in Tables 9.7 and 9.8. As seen from the tables, the permeation of ketoconazole from ketoconazole-fulvic acid complex (1:1) prepared by spray drying was found to be significantly higher as compared to ketoconazole alone or ketoconazole-fulvic acid (1:1) physical mixture. The difference was statistically significant ($p < 0.05$) using one-way repeated measure analysis of variance.

FUROSEMIDE PERMEABILITY STUDIES

In order to determine the effect of complexation on the permeability of furosemide, permeability studies for 1:2 freeze-dried complex of furosemide and fulvic acid were carried out by the rat everted gut sac method and compared with that of furosemide alone. For the study, everted rat intestinal sacs of about 5-cm length were prepared in tissue culture medium, TC199, and filled with about 3 mL of tissue culture medium. The sacs were placed in tubes containing 25 mL of tissue culture medium in which excess (equivalent to 40 mg equivalent of furosemide) of either furosemide alone or freeze-dried furosemide-fulvic acid (1:2) complex. The tubes were maintained in a shaking water bath at 37 °C, continuously bubbled with oxygen and agitated at a speed of 60 rpm.

Table 9.9: Concentration of furosemide in the serosal fluid at the start and after 1 hour during permeability studies by everted rat gut sac technique

Sample	Concentration of furosemide in the serosal fluid (µg/mL)	
	Initially	After 1 hour
Furosemide as such (FSM)	N.D.	0.497 ± 0.003
ITRA:FA (1:1) (freeze dried)	N.D.	5.014 ± 0.001

N.D.: Not detected.

The samples were withdrawn from the intestinal sac initially and after 1 hour. The samples were centrifuged for 5 min at 4000 rpm, filtered 0.22 µm membrane filter and analyzed by HPLC. The results of the study are shown in Table 9.9.

As seen from the results, the permeability of furosemide was observed to increase significantly (~10 times) across the rat everted gut sac on complexation with fulvic acid as compared to furosemide alone.

ANTIFUNGAL STUDIES

Itraconazole-fulvic Acid Complex

The minimum inhibitory concentration of itraconazole-fulvic acid complex prepared by spray drying was determined in comparison to the uncomplexed drug against a nonfilamentous fungus, *Candida albicans*, and a filamentous fungi, *Aspergillus fumigatus*, by the broth dilution method. The studies were carried out according to the macrodilution procedure of the National Committee for Clinical Laboratory Standards described in document M27-A2, Reference Method for Broth Dilution Susceptibility Testing of Yeasts (NCCLS, 2002a) for *Candida albicans* and according to document M38-A, Reference Method for Broth Dilution Susceptibility Testing of Filamentous Fungi (NCCLS, 2002b) for *Aspergillus fumigatus*.

Culture Medium

RPMI 1640 medium (with l-glutamine and phenol red, without bicarbonate) and buffered with 0.165 M morpholino propane sulfonic acid (MOPS) buffer at pH 7.0 was used as the culture medium.

For preparation of the medium, 10.4 g of powdered RPMI 1640 medium was dissolved in 900 mL of distilled water. To this was added 34.53 g of MOPS buffer and the pH was adjusted to 7.0 using 1 mol/L sodium hydroxide and the final volume was made up to 1000 mL with distilled water. The medium was filter sterilized using a 0.22 µm membrane filter and stored at 4 °C till further use.

Antifungal agents

Itraconazole alone and itraconazole-fulvic acid (1:1) complex prepared by spray drying were used as the antifungal agent. Fulvic acid alone was used as a control.

Organisms and Inoculum Preparation

Candida albicans

A clinical isolate of *C. albicans* (*C. albicans* VPCI 193) was obtained from Vallabhai Patel Chest Research Institute, New Delhi, and used for the study. Before use in the test, the isolate was subcultured twice on Sabauraud dextrose agar plates at 35 °C for 24 hours each. Five colonies of about 1 mm diameter from the 24 hour growth plates were picked and suspended in sterile saline (0.85%; w/v). The resulting suspension was vortexed for 15 seconds and the cell density was adjusted by adding sufficient sterile saline in order to match the transmittance to that produced by 0.5 McFarland turbidity standard ($BaSO_4$ turbidity standard) at 530 nm wavelength. This procedure yields a stock suspension of 1×10^6 to 5×10^6 cells/mL. A working suspension was prepared from this stock suspension by diluting 0.1 mL of stock suspension to 10 mL with RPMI 1640 medium followed by further dilution of 1 mL of the resulting suspension to 20 mL with RPMI 1640 medium.

Aspergillus fumigatus

A clinical isolate of *A. fumigatus* (*A. fumigatus* VPCI 68) was obtained from Vallabhai Patel Chest Research Institute, New Delhi, and used for the study. Before use in the test, the isolate was subcultured on potato dextrose agar plates at 35 °C for 7 days. Seven-day-old colonies were covered with approximately 1 ml of sterile saline (0.85%; w/v) containing 1% Tween 80 and the conidia were harvested by probing the colonies with the tip of the transfer pipette. The resulting mixture was transferred to a sterile tube and the heavy particles were allowed to settle for about 5 minutes. The upper homogeneous suspension was transferred to another sterile tube and vortexed for 15 seconds. The cell density was adjusted by adding sufficient sterile saline in order to obtain a transmittance of 80–82% for the stock suspension. A working suspension containing about 0.4×10^4 to 5×10^4 CFU/mL was prepared from this stock suspension by diluting 0.1 mL of stock suspension to 10 mL with RPMI 1640 medium.

Procedure

The tests were performed by following the standard additive twofold drug dilution scheme described in the NCCLS reference method. Stock solutions of itraconazole or itraconazole-fulvic acid complex were prepared at a concentration equivalent to 1600 µg/mL of itraconazole in DMSO. Stock solution of fulvic acid was prepared at a concentration of 2725 µg/mL which corresponded to the concentration of fulvic acid present in the itraconazole-fulvic acid complex. Dilutions in the range of 16-0.3125 µg/mL equivalent of itraconazole were prepared by additive dilution of the stock solutions with the culture medium, as described in the NCCLS method. Each drug dilution was then pipetted in 0.1-mL volumes into round-bottom, polystyrene, snap-cap, sterile tubes (12 × 75 mm) and inoculated by adding 0.9 mL volumes of the corresponding well-mixed working culture suspension of *C. albicans* or *A. fumigatus*. This step diluted each drug to the final test concentrations (16–0.03125 µg/mL equivalent of itraconazole). The growth control tube contained a 0.9-mL volume of inoculum suspension and a 0.1-mL volume of drug-free medium. Sterility control was performed by including 1-mL of un-inoculated, drug-free medium. The whole procedure was repeated thrice with each microorganism.

All tubes were incubated at 35 °C and observed for growth after 48 hours. MIC was visually determined as the lowest drug concentration which prevented any discernible growth.

The minimum inhibitory concentration of the itraconazole-fulvic acid complex prepared by spray drying was determined in comparison to the uncomplexed drug against *Candida albicans* and *Aspergillus fumigatus* by the broth dilution method. The results of the study are shown in Tables 9.10–9.13.

While itraconazole alone as well as the itraconazole-fulvic acid complex were able to inhibit the growth of *Candida albicans* as well as *Aspergillus fumigatus*, no inhibition was observed with

Table 9.10: Inhibition of the growth of *Candida albicans* by itraconazole and itraconazole-fulvic acid complex

Sl. No.	Conc. of itraconazole (μg/mL)	Growth of *Candida albicans* after 48 hours incubation								
		Itraconazole alone			Fulvic acid alone			Itraconazole-fulvic acid (1:1) spray dried		
		I	II	III	I	II	III	I	II	III
1.	Nil	Yes	Yes	Yes	Yes	Yes	Yes	Yes	Yes	Yes
2.	0.0313	Yes	Yes	Yes	Yes	Yes	Yes	Yes	Yes	Yes
3.	0.0625	Yes	Yes	Yes	Yes	Yes	Yes	Yes	No	No
4.	0.125	Yes	Yes	No	Yes	Yes	Yes	No	No	No
5.	0.25	No	No	No	Yes	Yes	Yes	No	No	No
6.	0.5	No	No	No	Yes	Yes	Yes	No	No	No
7.	1.0	No	No	No	Yes	Yes	Yes	No	No	No
8.	2.0	No	No	No	Yes	Yes	Yes	No	No	No
9.	4.0	No	No	No	Yes	Yes	Yes	No	No	No
10.	8.0	No	No	No	Yes	Yes	Yes	No	No	No
11.	16.0	No	No	No	Yes	Yes	Yes	No	No	No

Table 9.11: Minimum inhibitory concentration (μg/mL) of itraconazole against *Candida albicans*

Sl. No.	Sample	Minimum inhibitory concentration (MIC) against *Candida albicans* VPCI 193 (equivalent to μg/mL of itraconazole)
1.	Itraconazole powder (uncomplexed)	0.21 ± 0.07
2.	Itraconazole-fulvic acid (1:1) complex prepared by spray drying	0.08 ± 0.04

Table 9.12: Inhibition of the growth of *Aspergillus fumigatus* by itraconazole and itraconazole-fulvic acid complex

Sl. No.	Conc. of itraconazole (µg/mL)	Growth of *Aspergillus fumigatus* after 48 hours incubation								
		Itraconazole alone			Fulvic acid alone			Itraconazole-fulvic acid (1:1) spray dried		
		I	II	III	I	II	III	I	II	III
1.	Nil	Yes	Yes	Yes	Yes	Yes	Yes	Yes	Yes	Yes
2.	0.0313	Yes	Yes	Yes	Yes	Yes	Yes	Yes	Yes	Yes
3.	0.0625	Yes	Yes	Yes	Yes	Yes	Yes	No	Yes	Yes
4.	0.125	Yes	Yes	Yes	Yes	Yes	Yes	No	No	No
5.	0.25	Yes	Yes	No	Yes	Yes	Yes	No	No	No
6.	0.5	No	No	No	Yes	Yes	Yes	No	No	No
7.	1.0	No	No	No	Yes	Yes	Yes	No	No	No
8.	2.0	No	No	No	Yes	Yes	Yes	No	No	No
9.	4.0	No	No	No	Yes	Yes	Yes	No	No	No
10.	8.0	No	No	No	Yes	Yes	Yes	No	No	No
11.	16.0	No	No	No	Yes	Yes	Yes	No	No	No

Table 9.13: Minimum inhibitory concentration (µg/mL) of itraconazole against *Aspergillus fumigatus*

Sl. No.	Sample	Minimum inhibitory concentration (MIC) against *Aspergillus fumigatus* VPCI 68 (equivalent to µg/mL of itraconazole)
1.	Itraconazole powder (uncomplexed)	0.42 ± 0.14
2.	Itraconazole-fulvic acid (1:1) complex prepared by spray drying	0.10 ± 0.04

fulvic acid alone indicating that it does not have any anti-fungal activity of its own. The minimum inhibitory concentration (MIC) of itraconazole alone for both Candida as well as Aspergillus was found to be within the reported range (NCCLS, 2002a,b).

The mean MIC values for itraconazole-fulvic acid complex for both Candida as well as Aspergillus were lower than itraconazole alone indicating that the prepared complex has an antifungal activity which is greater than uncomplexed drug. The difference was statistically significant ($p < 0.05$) using one-way repeated measure analysis of variance. The lower minimum inhibitory concentration attained with the itraconazole-fulvic acid complex could be due to better permeability of the complex in comparison to uncomplexed drug.

KETOCONAZOLE-FULVIC ACID COMPLEX

The minimum inhibitory concentration of the ketoconazole-fulvic acid complex prepared by spray drying was determined in comparison to the ketoconazole alone against *Candida albicans* by the serial broth dilution technique in a manner similar to that described for itraconazole. The results of the study are shown in Tables 9.14 and 9.15.

Table 9.14: Inhibition of the growth of *Candida albicans* by ketoconazole and ketoconazole-fulvic acid complex

Sl. No.	Conc. of ketoconazole (µg/mL)	Growth of *Candida albicans* after 48 hours incubation								
		Ketoconazole alone			Fulvic acid alone			Ketoconazole-fulvic acid (1:1) spray-dried complex		
		I	II	III	I	II	III	I	II	III
1	NIL	Yes	Yes	Yes	Yes	Yes	Yes	Yes	Yes	Yes
2	0.0313	Yes	Yes	Yes	Yes	Yes	Yes	Yes	Yes	Yes
3	0.0625	Yes	Yes	Yes	Yes	Yes	Yes	Yes	Yes	Yes
4	0.125	Yes	Yes	Yes	Yes	Yes	Yes	Yes	Yes	Yes
5	0.25	Yes	Yes	Yes	Yes	Yes	Yes	Yes	Yes	Yes
6	0.5	Yes	Yes	Yes	Yes	Yes	Yes	Yes	Yes	Yes

Table 9.14: Continued ....

S. No.	Conc. of ketoconazole (µg/mL)	Growth of *Candida albicans* after 48 hours incubation								
		Ketoconazole alone			Fulvic acid alone			Ketoconazole-fulvic acid (1:1) spray-dried complex		
		I	II	III	I	II	III	I	II	III
7	1.0	Yes	Yes	Yes	Yes	Yes	Yes	Yes	Yes	Yes
8	2.0	Yes	Yes	Yes	Yes	Yes	Yes	No	No	No
9	4.0	Yes	Yes	Yes	Yes	Yes	Yes	No	No	No
10	8.0	No	No	No	Yes	Yes	Yes	No	No	No
11	16.0	No	No	No	Yes	Yes	Yes	No	No	No

Table 9.15: Minimum inhibitory concentration (µg/mL) of ketoconazole against *Candida albicans*

Sl. No.	Sample	Minimum inhibitory concentration (MIC) against *Candida albicans* (equivalent to µg/mL of ketoconazole)
1.	Ketoconazole as such	8.0 ± 0.0
2.	Ketoconazole-fulvic acid complex (1:1) spray dried complex	2.0 ± 0.0

Ketoconazole alone as well as the ketoconazole-fulvic acid complex were able to inhibit the growth of candida albicans while no inhibition was observed with fulvic acid alone indicating that it does not have any antifungal activity of its own. The mean MIC values for ketoconazole-fulvic acid complex for *Candida albicans* was lower than ketocanazole alone indicating that the prepared complex has an antifungal activity which is four times greater than uncomplexed drug. The lower minimum inhibitory concentration attained with the ketoconazole-fulvic acid complex could be due to better solubility and pemeability of complex in comparison to uncomplexed drug.

10

BIOAVAILABILITY ENHANCING PROPERTIES OF HUMIC SUBSTANCES

Bioavailability is a pharmacokinetic term that describes the rate and extent to which the active drug ingredient is absorbed from a drug product and becomes available at the site of drug action. Since pharmacologic response is generally related to the concentration of drug at the receptor site, the availability of a drug from a dosage form is a critical element of a drug product's clinical efficacy. However, drug concentrations usually cannot be readily measured directly at the site of action. Therefore, most bioavailability studies involve the determination of drug concentration in the blood or urine. This is based on the premise that the drug at the site of action is in equilibrium with drug in the blood. It is therefore possible to obtain an indirect measure of drug response by monitoring drug levels in the blood or urine. Thus, bioavailability is concerned with how quickly and how much of a drug appears in the blood after a specific dose is administered. The bioavailability of a drug product often determines the therapeutic efficacy of that product since it affects the onset, intensity, and duration of therapeutic response of the drug. In most cases, one is concerned with the extent of absorption of drug, (that is, the fraction of the dose that actually reaches the bloodstream) since this represents the "effective dose" of a drug. This is generally less than the amount of drug actually administered in the dosage form. In some cases, notably those where acute conditions are being treated, one is also concerned with the rate of absorption of a drug, since rapid onset of pharmacologic action is desired. Conversely, there are instances where a slower rate of absorption is desired, either to avoid adverse effects or to produce a prolonged duration of action.

The authors and coworkers investigated the bioavailability-enhancing properties of humic substances by carrying out comparative bioavailability and thermodynamic studies of formulation prepared using complexes of various drugs with humic substances with that of formulations containing uncomplexed drugs.

BIOAVAILABILITY-ENHANCEMENT OF ITRACONAZOLE

Khanna (2005) carried out a comparative single-dose oral bioavailability study of tablets containing spray-dried itraconazole-fulvic acid complex and the innovator's formulation of itraconazole (Sporanox capsule 100 mg, M/s Janssen Pharmaceutica, USA) in six healthy human volunteers in the fasting state.

Study Design

The study was conducted as an open randomized, two treatment, two sequence, two period, single dose, crossover, comparative bioavailability study on itraconazole formulations comparing itraconazole 100 mg tablet prepared using itraconazole-fulvic acid complex with Sporanox 100 mg capsules of Janssen Pharmaceutica, USA, in healthy, adult, male, human subjects under fasting conditions.

The subjects were admitted and housed in the Clinical Pharmacology Unit from at least 10–12 hours before dose administration and were discharged 24 hours after administration of the test or reference products during each period. After discharge at 24 hours, subjects made two ambulatory visits to the Clinical Pharmacology Unit for blood sampling at 48 and 72 hours.

A single oral dose of either the reference (Sporanox capsule 100 mg) or test product (itraconazole tablet containing itraconazole-fulvic acid complex equivalent to 100 mg of itraconazole) was administered with 240 mL of drinking water under the supervision of a trained Medical Officer.

All subjects were required to fast overnight after admission for at least 10 hours before the morning dose and for 4 hours postdose. The subjects received standard meals, *i.e.*, lunch, snacks, and dinner at approximately 4, 9, and 13 hours respectively, after first dosing. During housing, all meal plans were identical for the two periods. In case where the meals and blood sample collection time coincided, samples were collected before meals were provided.

Drinking water was not allowed from 1 hour before dosing and until 2 hours postdose. Thereafter, it was allowed at all times.

Sampling Schedule

A total of 36 Five-milliliter blood samples for both the treatments were collected in EDTA vacutainers during the course of the study through an indwelling cannula placed in the forearm vein. The blood samples were collected pre-dose and at 0.5, 1, 1.5, 2, 2.5, 3, 3.5, 4, 4.5, 5, 5.5, 6, 8, 12, 24, 48, and 72 hours after each oral dose in each period. There was a washout period of 14 days between the administration of test and reference products.

Restrictions

Only subjects who had not received any medication including over the counter (OTC) medications during the 2 weeks period prior to the onset of the study were recruited. They were instructed during screening not to take any prescription and OTC medications subsequently until the completion of the study.

All subjects were instructed to abstain from any alcoholic product for 48 hours prior to dosing until completion of the study. They also abstained from any xanthine-containing food or beverages during in-house stay in each period.

Selection of Subjects

Adequate numbers of subjects were selected randomly and were subjected to a standardized screening procedure. Six healthy male subjects were selected from the screened ones on the following inclusion and exclusion criteria:

Inclusion Criteria

1. Be in the age range of 18–45 yrs.
2. Be neither overweight nor underweight for his height.
3. Have voluntarily given written informed consent to participate in this study.
4. Be of normal health as determined by the medical history and physical examination of the subjects performed within 28 days prior to the commencement of the study.

Exclusion Criteria

1. History of allergy to itraconazole and/or related drugs.
2. Any evidence of organ dysfunction or any clinically significant deviation from the normal, in physical or clinical determinations.
3. Presence of disease markers of HIV 1 and 2, Hepatitis B and C viruses and syphilis infection.
4. Presence of values which are clinically significantly different from normal reference ranges for hemoglobin, total white blood cells count, differential WBC count, and platelet count.
5. Positive for urinary screen testing of drugs of abuse (opiates and cannabinoids).
6. Presence of values which are significantly different from normal reference ranges for serum creatinine, blood urea, serum aspartate, aminotransferase (AST), serum alanine aminotransferase (ALT), serum alkaline phosphatase, serum bilirubin, plasma glucose, and serum cholesterol.
7. Clinically abnormal chemical and microscopic examination of urine defined as presence of RBC, WBC (> 4/HPF), epithelia cells (> 4/HPF), glucose (positive), and protein (positive).
8. Clinically abnormal ECG and chest X-ray.
9. History of serious gastrointestinal, hepatic, renal, cardiovascular, pulmonary, neurological or hematological disease, diabetes, or glaucoma.

10. History of any psychiatric illness, which may impair the ability to provide, written informed consent.
11. Regular smokers who smoke more than 10 cigarettes daily or have difficulty abstaining from smoking for the duration of each study period.
12. History of drug dependence or excessive alcohol intake on a habitual basis of more than two units of alcoholic beverages per day (one unit equivalent to half pint of beer or one glass of wine or one measure of spirit) or have difficulty in abstaining for the duration of each study period.
13. Use of any enzyme modifying drugs within 30 days prior to day 1 of this study.
14. Participation in any clinical trial within 12 weeks preceding day 1 of this study.

Safety

Vital signs of oral temp, sitting blood pressure, and radial pulse were measured during subject admission prior to dosing and 2, 8, and 12 hours after administration of study drug and before discharge in each period. Clinical examination of the subjects was conducted by a qualified medical designate on duty after subject admission prior to dosing of study drug and before discharge.

Ethical Consideration

The study was carried out as per ICH (Step 5), "Guidance for Good Clinical Practice" and the principles enunciated in the Declaration of Helsinki. The protocol and the corresponding informed consent form used to obtain informed consent of the study subjects was reviewed and approved by the Institutional Review Board.

Analysis of Plasma Samples

A liquid-liquid extraction method followed by HPLC was used for the analysis of itraconazole in plasma. The method was based on the methods reported by Woestenborghs *et al.* (1987) and Warnock *et al.* (1988) with slight modification. Statistical and pharmacokinetic parameters were calculated using the WinNonlin Pharmacokinetic software. ANOVA and correlation analysis was applied for pharmacokinetic parameters.

Table 10.1 and Figure 10.1 show the mean plasma levels of itraconazole obtained for the reference as well as the test preparation. The different pharmacokinetic parameters calculated

Table 10.1: Mean plasma concentration of itraconazole achieved for reference and test formulatios

Time (hr)	Mean (±SD) plasma concentration of itraconazole (ng/mL)	
	Reference product (Sporanx capsules 100 mg)	Test product (Itraconazole-fulvic acid tablet)
0	0	0
0.5	2.39 ± 1.31	10.85 ± 4.01
1	6.52 ± 3.11	23.83 ± 11.71
1.5	15.16 ± 7.01	35.00 ± 18.37
2	21.38 ± 7.01	42.55 ± 12.21
2.5	26.36 ± 7.17	53.67 ± 7.66
3	31.37 ± 6.54	59.22 ± 4.71
3.5	36.47 ± 6.98	61.17 ± 10.13
4	35.49 ± 7.02	55.93 ± 10.21
4.5	32.77 ± 7.95	51.42 ± 9.34
5	30.98 ± 9.93	45.89 ± 8.23
5.5	27.38 ± 8.29	41.47 ± 9.50
6	24.58 ± 6.90	35.98 ± 10.01
8	20.06 ± 6.10	29.10 ± 7.47
12	16.18 ± 6.30	23.29 ± 4.72
24	9.10 ± 3.90	15.07 ± 3.18
48	4.20 ± 1.44	9.25 ± 2.19
72	1.34 ± 1.13	4.84 ± 1.46

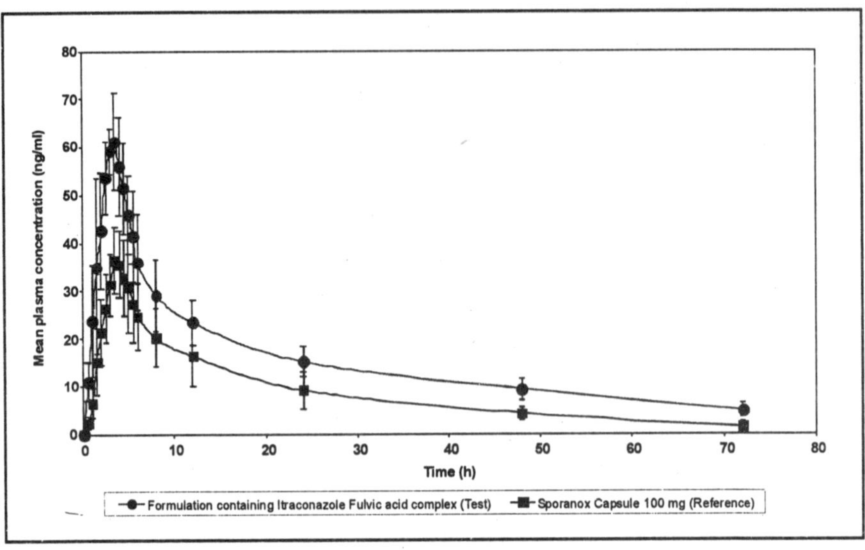

Fig. 10.1: Comparison of the mean plasma profile of formulation containing itraconazole-fulvic acid complex prepared by spray drying (Test) with that of marketed formulation (Sporanox capsules) (Reference) in six healthy human volunteers

for both the formulations are shown in Table 10.2. The calculated parameters were analyzed by ANOVA single factor analysis and were found to be significant at 5% confidence level.

As can be seen from Table 10.2, the mean value of C_{max} for the developed formulation of itraconazole (66.82 ± 7.35 ng/mL) was significantly higher than that of the marketed formulation (40.29 ± 8.75 ng/mL). Similarly, the mean values of $AUC_{0\text{-}72}$ and $AUC_{0\text{-}\infty}$ for the developed formulation were also found to be higher than that of the marketed formulation. The differences were found to be statistically significant ($p = 0.002$). There was no significant difference in the other parameters (T_{max}, K_{el}, and $T_{1/2}$) for the two products.

The results of the study clearly indicated that complexation of itraconazole with fulvic acid significantly increased its bioavailability in humans in comparison to uncomplexed drug formulation.

Table 10.2: Comparison of the mean pharmacokinetic parameters for sporanox capsules 100 mg (reference formulation) and itraconazole-fulvic acid tablet 100 mg (test formulation)

Pharmacokinetic parameters in 6 healthy human volunteers	Sporanox capsule 100 mg (M/s Janssen Pharmaceutica, USA) (Reference)	Itraconazole tablet 100 mg containing itraconazole-fulvic acid complex (Test)
Cmax (ng/mL)	40.29 ± 8.75	66.82 ± 7.35
Tmax (hr)	3.83 ± 0.68	2.83 ± 0.75
$AUC_{0\text{-}72}$ (nghr/mL)	634.29 ± 204.81	1110.49 ± 202.70
$AUC_{0\text{-}\infty}$ (nghr/mL)	709.03 ± 220.38	1276.01 ± 259.88
Kel hr^{-1}	0.033 ± 0.007	0.031 ± 0.005
T1/2 (hr)	22.06 ± 4.86	23.08 ± 3.11

BIOAVAILABILITY-ENHANCEMENT OF KETOCONAZOLE

Karmarkar (2007) studied the oral bioavailability of the tablets containing spray dried ketoconazole-fulvic acid complex equivalent to 200 mg ketoconazole in comparison to that of marketed ketoconazole tablet (Nizrol of Jenssen pharmacautica, India) in six healthy male human volunteers under fasting condition.

Study Design

The study was an open randomized, two treatment, two sequence, two period, single dose, crossover, comparative bioavailability study on ketoconazole formulations comparing Ketoconazole 200 mg tablet prepared using Ketoconazole-fulvic acid complex with Nizrol 200 mg tablet of Janssen Pharmaceutica, India, in healthy, adult, male, human subjects under fasting conditions.

The subjects were admitted to the Clinical Pharmacology Unit from at least 10–12 hours before dose administration and were discharged 24 hours after administration of the test or reference products during each period. After discharge at 24 hours, subjects made one ambulatory visit for blood sampling at 48 hours.

A single oral dose of either the reference (Nizrol tablets 200 mg) or test product (ketoconazole tablet containing ketoconazole-fulvic acid complex equivalent to 200 mg of ketoconazole) was administered with 240 mL of drinking water under the supervision of a trained Medical Officer.

All subjects are required to fast overnight after admission for at least 10 hours before the morning dose and for 4 hours postdose. The subjects received a standard meal, *i.e.*, lunch, snacks, dinner at approximately 4, 9, and 13 hours respectively, after first dosing. Drinking water was not allowed from 1 hour before dosing until 2 hours postdose. Thereafter, it was allowed at all times.

Sampling Schedule

A total of 30 Five-milliliter blood samples for both the treatments were collected in EDTA vacutaners during the course of the study through an indwelling cannula placed in a forearm vein. The blood samples were collected predose and at 0.5, 1, 1.5, 2, 2.5, 3, 3.5, 4, 5, 6, 8, 12, 24, and 48 hours after each oral dose in each period. There was a washout period of at least 4 days between the administration of test and reference products.

The inclusion and exclusion criteria as well as other ethical and safety considerations were similar to the comparative bioavailability study of itraconazole described above.

Analysis of Plasma Samples

A liquid-liquid extraction method followed by HPLC was used for the analysis of ketoconazole in plasma. The method was based on the methods reported by Woestenborghs *et al.* (1987) and Warnock *et al.* (1988) with slight modification. Statistical and pharmacokinetic parameters were calculated using the WinNonlin Pharmacokinetic software. ANOVA and correlation analysis was applied for pharmacokinetic parameters.

Table 10.3 and Figure 10.2 show the mean plasma levels of ketoconazole obtained for the reference as well as the test preparation. The different pharmacokinetic parameters calculated

Table 10.3: Mean plasma concentration of itraconazole achieved for reference and test formulatios

Time (hr)	Mean (±SD) plasma concentration of ketoconazole (ng/mL)	
	Reference product (Nizrol tablet 200 mg)	Test product (Ketoconazole-fulvic acid tablet)
0	0	0
0.5	0.58 ± 0.14	1.00 ± 0.12
1	1.11 ± 0.47	2.62 ± 0.18
1.5	2.26 ± 0.22	3.94 ± 0.40
2	1.44 ± 0.14	2.53 ± 0.24
2.5	0.96 ± 0.05	1.88 ± 0.23
3	0.80 ± 0.17	1.69 ± 0.17
3.5	0.72 ± 0.16	1.59 ± 0.19
4	0.48 ± 0.09	0.90 ± 0.08
5	0.30 ± 0.08	0.56 ± 0.10
6	0.20 ± 0.08	0.29 ± 0.09
8	0.09 ± 0.02	0.12 ± 0.03
12	0.02 ± 0.02	0.06 ± 0.01
24	0.009 ± 0.00	0.01 ± 0.016
48	0.001 ± 0.00	0.004 ± 0.002

for both the formulations are shown in Table 10.4. The calculated parameters were analyzed by ANOVA single factor analysis and were found to be significant at 5% confidence level.

As can be seen from Table 10.4, the mean value of C_{max} for the developed formulation of ketoconazole (3.945 ± 0.406 mg/mL) was significantly higher than that of the marketed formulation (2.268 ± 0.298 mg/mL). Similarly, the mean values

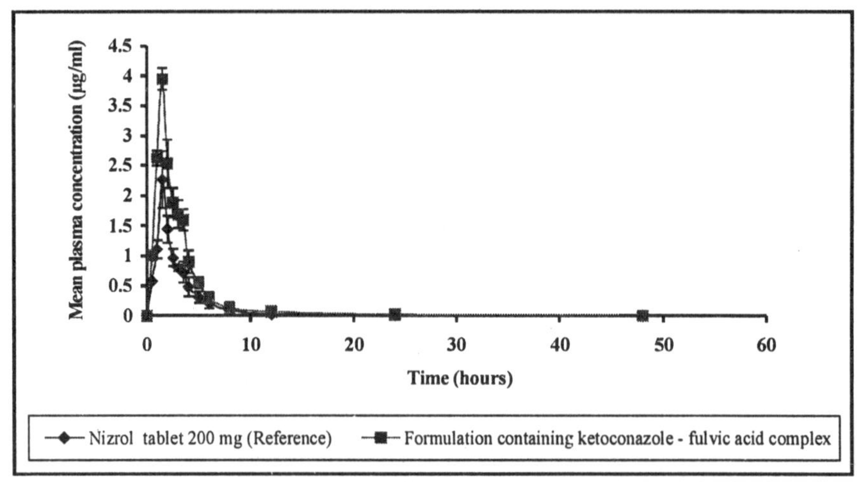

Fig. 10.2: Comparison of the mean plasma profile of formulation containing ketoconazole-fulvic acid (1:1) complex prepared by spray drying (Test) with that of marketed formulation (Nizrol tablet) (Reference) in six healthy human volunteers

Table 10.4: Comparison of the mean pharmacokinetic parameters for Nizrol, M/s Janssen Pharmaceutica, India, tablet 200 mg (reference formulation) and ketoconazole-fulvic acid (1:1) spray-dried complex tablet 200 mg (test formulation)

Pharmacokinetic parameters in 6 healthy human volunteers	Nizrol tablet 200 mg (M/s Janssen Pharmaceutica, India) (Reference)	Ketoconazole tablet 200 mg containing ketoconazole-fulvic acid complex (Test)
C_{max} (mg/mL)	2.26 ± 0.298	3.94 ± 0.406
T_{max} (hr)	1.5 ± 0.0	1.50 ± 0.00
AUC_{0-48} (mg.hr/mL)	5.49 ± 0.680	10.38 ± 0.7907
$AUC_{0-\infty}$ (mg.hr/mL)	5.51 ± 0.679	10.43 ± 0.817
$K_{el(\alpha)}(hr^{-1})$	0.421 ± 0.046	0.370 ± 0.039
$T_{1/2(\alpha)}(hr)$	1.65 ± 0.175	1.881 ± 0.18
$K_{el(\beta)}(hr^{-1})$	0.092 ± 0.005	0.082 ± 0.005
$T_{1/2(\beta)}(hr)$	7.49 ± 0.39	8.39 ± 0.56

of AUC_{0-48} and $AUC_{0-\infty}$ for the developed formulation were also found to be higher than that of the marketed formulation. The differences were found to be statistically significant ($p < 0.01$). There was no significant difference in the other parameters (T_{max}, K_{el}, and $T_{1/2}$) for the two products.

The results of the study clearly indicated that complexation of ketoconazole with fulvic acid significantly increased its bioavailability in humans in comparison to uncomplexed drug formulation.

SAFETY AND TOXICITY STUDIES ON HUMIC SUBSTANCES

The long traditional and documented use of humic substances in folklore medicine for thousands of years without any reports or incidences of toxicity is in itself a testimony to their safety for human use. Yet, a number of studies have been carried out worldwide, in the last 50 years, to document its safety based on currently accepted protocols for toxicity and safety studies.

ACUTE AND CHRONIC TOXICITY STUDIES ON SHILAJIT

Assiri *et al.* (2003) carried out acute (24 hours) and chronic (90 days) toxicity studies of shilajit in mice.

Specimen

Shilajit was collected from mountains of district Zoab of Balochistan province in Pakistan and stored at 4 °C. The fresh aqueous suspension was prepared in distilled water before administration.

Animal stock

Swiss albino mice (home bred) aged 6–7 weeks, weighing about 24–28 g, and fed on Purina Chow diet and water ad libitum were used in the study. The animals were maintained under controlled temperature, humidity, and automated light cycles (12 hr light/ 12 hr dark).

Acute toxicity

A total of 20 mice were randomly allotted to one control and three treated groups. The extract in each case was administered orally in three doses, namely, 0.5 g/kg, 1.0 g/kg, and 3 g/kg. The animals were observed for 24 hours for all signs of toxicity and mortality.

Chronic toxicity

A total of 20 male and 20 female mice were randomly allotted to the control and test groups. The extract in each case was given in drinking water. The dose selected was 100 mg/kg/day, which is 1/5 of the pharmacologically active dose commonly used for crude drugs. The treatment was continued for a period of 3 months. The animals were observed for all external general symptoms of toxicity, body weight changes, and mortality. The average pre- and post-treatment body weights, vital organ weights, and viscera of the chronically treated animals were compared with the control group. The chronically treated male animals were also analyzed for spermatogenic dysfunction using the sperm abnormality test, which is considered to be a reliable parameter for assessing germ cell mutagenicity and carcinogenicity. The caudae epididymides and vas deferens from the same animals were dissected out and transferred to a centrifuge tube containing 3 mL Krebs Ringer bicarbonate buffer. The sperm suspension was filtered through an 80 urn silk mesh to remove tissue fragments and 0.5 mL of 1% eosin Y was added to each tube. The contents were thoroughly mixed and the slides were made by placing one drop of Ae solution on a slide and spread by three passes of another slide. Coded slides were examined for the sperm abnormalities.

Biochemical studies

The blood was collected from chronically treated animals directly by heart puncture method under anesthesia. The serum was separated and stored at -20 °C until analyzed for alanine aminotransferase (ALT), aspartate aminotransferase (AST), creatine kinase isoenzyme MB (CK–MB), glucose, urea, and creatinine. These parameters were analyzed on a spectrophotometer (Ultrospec II, LKB) by using test combination reagents (Boehringer Mannheim GmbH, Diagnostical, Germany).

Haematological studies

The blood was analyzed for WBC, RBC, and hemoglobin using Contraves Digicell 3100H (Zurich).

Statistical analysis

The data collected for different parameters was subjected to statistical analyses and the means were compared using the Chi-square test or Student's *t*-test.

Results

In acute toxicity test, shilajit exhibited no visible signs of toxicity, although some central nervous system (CNS) stimulation was noticed at a dose of 3 g/kg. No mortality was observed up to 3 g/kg dose level. The CNS stimulation at higher dose observed in shilajit-treated mice during our study may be attributed to its effects on brain monoamines.

No toxicity symptoms were observed in mice upon chronic treatment with Shilajit. All the treated male and female mice throughout the study remained normal and were comparable to the control. The weight gain by animals in shilajit-treated group was significant ($p < 0.001$) as was also observed in the control group. At the end of chronic treatment, the average weights of vital organs and condition of the viscera were normal and comparable to the control. However, the weight of testes, caudae epididymides, and seminal vesicles was found to be significantly higher ($p < 0.05$) in treated group as compared to that of the control. The increase in weight of testes has been attributed to different chemical constituents of natural drugs possessing androgenic activities. The results indicated that Shilajit treatment did not induce any significant change ($p > 0.05$) in the biochemical parameters as compared to the control.

The hematological studies revealed a significant rise of red blood cells (RBC) and hemoglobin levels in shilajit-treated animals as compared to that of the control. The altered red cell production may be attributed to the constituents of natural drugs that may influence the androgen level in the body.

The effect of shilajit on the development of mice embryo was also studied. A total of 71 pregnant female mice were given shilajit (250 and 500 mg/kg) orally via needle tube, daily from day 8–12 of pregnancy. All the treated and control animals showed no differences in the number of the litter size, the placenta, and the body weight of the embryos and the number of resorped embryos at day 17 of gestation. However, few abnormalities were observed in both treated and control groups (Ahmed *et al.*, 2003).

Shilajit extract did not cause any mortality in mice up to the dose of 1 g/kg (intraperitoneal injection) (Acharya *et al.*, 1988). For toxicological study, the experimental animals received the preparation daily in the form of 1-10% aqueous solution (orally) for 1 month. The daily doses of shilajit extract for rabbits and mice were 0.05, 0.1, 0.15, 0.2, 0.3, 0.4, and 0.5 g/kg. On its application both once (0.5 g/kg) and on a multitime basis (total dose was from 1.5 to 15 g/kg) the investigators did not observe any morphological or histological changes in the internal organs of animals in comparison with the control group (Scheptekin *et al.*, 2002).

In the Ukrainian Gerontology Institute, the study of toxicological properties of mumie picked from alpine regions of Central Asia was carried out. It was found that application of the remedy at the doses of 0.2 and 1 g/kg for 3 months did not lead to negative influence on the function of heart, liver, kidneys, blood cells, or nervous and endocrine systems. The study of specific teratogenic action showed that treatment of pregnant rats with mumie did not render embryotoxic or teratogenic actions. The postnatal development of young rats, whose parents received the preparation, was also normal.

Most of the investigators noted absence of side effects with mumie application at daily dose of 0.1–0.3 g inwardly. Some patients with bone fractures felt burning in the region of fracture. Patients with chronic colitis felt heat, burning, weakness, and sweating during 40–60 minutes after application of mumie extract. At higher doses (0.9–1.5 g/day), it can lead to increase in body temperature to 37.5 °C, sweating, and headache. The duration of this reaction was from 20 minutes to 2–3 hours (Schepetkin *et al.*, 2002).

Ghosal *et al.* (1991) reported that the LD_{50} of FAs (containing DBPs to the extent of 3–8 %; w/w) and MCB by the oral route, in albino rats, was 1268 ± 78 mg/kg and 684 ± 32 mg/kg, respectively. These data suggest that the acute toxicity of both FAs and MCB is of a considerably low order and that there is a wide margin between the effective and the acute toxic doses.

Extensive medical studies have shown that the toxicity of naturally occurring fulvic acid is remarkably low. Fulvic acid has been approved for marketing by the USFDA and other regulatory bodies of Europe and Asia as component of dietary supplements and by the Chinese drug authorities for pharmaceutical use. Table 11.1 lists the various marketed products containing fulvic acid.

Table 11.1: Some of the products containing humic substances marketed worldwide

Product	Manufacturer	Countries where marketed
Humet R Syrup	Humet Corporation	USA, UK, Taiwan, Russia, Portugal, The Netherlands
Humet R Capsules	Humet Corporation	USA, UK, Taiwan, Russia, Portugal, The Netherlands
Humifulvate Capsules	Humet Corporation	USA, UK, Taiwan, Russia, Portugal, The Netherlands
Humetta Effervescent Tablets	Humet Corporation	USA, UK, Taiwan, Russia, Portugal, The Netherlands
Maximol solution	Neways International	USA, UK
Silverzone Nasal spray	Gifts of Nature Inc.	USA, UK
Immunite solution	Westwood Enterprises	UK

SUBACUTE TOXICITY STUDIES OF FULVIC ACID AND FULVIC ACID DRUG COMPLEXES

Khanna (2005) carried out 21 days subacute toxicity studies on fulvic acid and fulvic acid-itraconazole complexes using rat as the model animal.

Type of animals used

Wistar rats weighing between 150 and 180 g were used in the study.

Number of animals used

Five animals were used in each group, *i.e.*, control, placebo, and test groups.

Procedure

The animals were divided into three groups: control, placebo and test group. The test group was given the formulation containing itraconazole-fulvic acid complex equivalent to 30 mg/kg body weight, suspended in water. This corresponds to about 10 times the average recommended daily dose for humans. The placebo group was administered the complexing agent, *i.e.*, fulvic acid solution in distilled water, in the dose of 50 mg/kg body weight. This amount corresponded to the amount of fulvic acid present in the test formulation being administered. The control group was given the vehicle alone.

Before starting the experiment, about 2.0 mL blood was collected from the tail vein of each rat from all the three groups. The blood samples were analyzed for hematological parameters, kidney function test, and liver function test.

The three groups were then administered the respective products daily for 21 days, with the help of a mouth feeder, and observed for physical activity and body weight. After a period of 21 days, blood samples were again withdrawn from the rats and tested for hematological parameters, liver function test, and kidney function test.

After giving the test medication for a period of 21 days, the control, placebo, and test animals were sacrificed. Liver, kidney, heart, and spleen were taken out of the rats and stored in 10% formalin solution. The tissues were washed with normal saline and were seen for:

Gross Changes

Color changes or the development of any patch was observed.

Histopathological Changes

Tissues were cut into thin slices and kept in 10% formalin solution for fixing for a period of 48 hours. This prevented the postmortem changes such as putrefaction and autolysis and preserved the cell constituents in as life-like manner as possible. It protected the tissues by hardening the naturally soft tissues thereby allowing easy manipulation during subsequent processing. Slides were prepared and observed under the microscope.

The following results were obtained:

Physical activity

No abnormality in the physical activity of the rats was observed in any of groups.

Body weight

Normal increase in the body weight of all the rats was observed in all the groups. The results are shown in Table 11.2.

Table 11.2: Body weight of rats

Treatment	**Mean (± SD; n=5) body weight of rats**			
	0 day	**After 7 days**	**After 14 days**	**After 21 days**
Control	169 ± 10.8	194 ± 8.9	210 ± 7.9	230 ± 6.1
Placebo	167 ± 7.5	196 ± 9.6	212 ± 10.3	223 ± 10.3
Test	174 ± 6.5	199 ± 9.6	213 ± 5.7	226 ± 7.4

Hematological Parameters

The hematological parameters, *viz.*, hemoglobin, total leucocyte count, and red blood cell count for the test formulation were not significantly different from the initial values as well as from the control group after 21 days. The results are shown in Table 11.3.

Table 11.3: Hematological test values

Treatment	Mean (± SD; n=5) levels in the blood					
	0 day			After 21 days		
	HB (g/dL)	TLC (×10^3/mm^3)	RBC (×10^3/mm^3)	HB (g/dL)	TLC (×103/mm^3)	RBC (×10^3/mm^3)
Control	13.8 ± 0.6	18300 ± 4549	6.90 ± 0.38	14.4 ± 0.4	13840 ± 288	5.60 ± 0.30
Placebo	13.9 ± 0.9	12480 ± 2654	7.12 ± 0.38	14.6 ± 0.6	14160 ± 691	5.92 ± 0.38
Test	13.7 ± 1.3	18625 ± 4242	6.82 ± 0.68	13.9 ± 0.6	13560 ± 757	5.30 ± 0.32

HB, Hemoglobin; TLC, total leucocyte count; RBC, red blood cells.

Table 11.4: Kidney function test values

Treatment	Mean (± SD; n=5) levels in the blood			
	0 day		After 21 days	
	Urea (mg%)	Creatinine (mg/dL)	Urea (mg%)	Creatinine (mg/dL)
Control	40.6 ± 4.5	1.16 ± 0.16	39.8 ± 4.4	1.10 ± 0.15
Placebo	35.2 ± 2.6	1.24 ± 0.11	38.6 ± 3.2	1.20 ± 0.12
Test	52.4 ± 33.8	1.14 ± 0.08	43.8 ± 5.4	1.26 ± 0.13

Table 11.5: Liver function test values

Treatment	Mean (± SD; n=5) levels in the blood							
	0 days				After 21 days			
	Bil (mg%)	SGOT (IU/L)	SGPT (IU/L)	A.PHOS (IU/L)	Bil (mg%)	SGOT (IU/L)	SGPT (IU/L)	A.PHOS (IU/L)
Control	0.38 ± 0.08	306 ± 98.2	57.8 ± 8.1	216 ± 122	0.52 ± 0.19	162 ± 32	42.6 ± 4.3	232 ± 24
Placebo	0.36 ± 0.11	231 ± 37.4	48.6 ± 4.7	417 ± 161	0.50 ± 0.14	257 ± 135	46.6 ± 5.3	196 ± 35
Test	0.44 ± 0.11	213 ± 44.1	49.4 ± 9.1	287 ± 101	0.48 ± 0.08	190 ± 90	43.4 ± 5.6	221 ± 45

Bil: total bilirubin; A. Phos.: alkaline phosphatase; SGOT: serum glutamate oxaloacetate transaminase; SGPT: serum glutamate pyruvate transaminasea

Kidney Function Test

The parameters indicative of kidney function, *viz.*, urea and creatinine for the test formulation were not significantly different from the initial values as well as from the control group after 21 days. The results are shown in Table 11.4.

Liver Function Test

The parameters indicative of liver function, *viz.*, bilirubin, SGOT, SGPT, and alkaline phosphatase for the test formulation were not significantly different from the initial values as well as from the control group after 21 days. The results are shown in Table 11.5.

Histopathological Studies

Liver

Control : Normal architecture with central vein and portal traits were seen.

Placebo : Normal appearance.

Test : No changes were seen. Architecture was normal with central vein.

Kidney

Control : Multiple sections were examined from both left and right kidney. Cortex and medulla consisting of variable sizes were seen. Normal cellularities were seen. Renal tubules showed normal histology.

Placebo : Normal appearance.

Test : No changes were observed.

Heart

Control : Muscle fibers arranged in bundles were seen.

Placebo : Normal appearance.

Test : Normal with muscle fibers arranged in bundles were seen.

Spleen

Control : Lymphoid follicles with sinusoids filled with blood were seen.

Placebo : Normal histology was seen.

Test : No changes were observed.

CONCLUSION

No signs of toxicity were observed in rats when either fulvic acid or the test formulation containing itraconazole-fulvic acid was given at 10 times the average normal dose of itraconazole for 21 days, thereby confirming their safety.

12

HUMIC SUBSTANCES IN DRUG DEVELOPMENT

Permeability and solubility are key parameters which determine the fate of an orally administered drug dosage form in the gastro-intestinal tract. Molecules with too-low permeability and/or solubility usually provide a low and variable bioavailability and present a significant challenge to the pharmaceutical scientist to formulate them into highly bioavailable and therapeutically useful products. A number of new drugs discovered do not get marketed because of their poor biopharmaceutical characteristics. Recent research on humic substances has shown that these can be employed as useful tools for solving such biopharmaceutical problems. A few categories of modern drugs where the technology can find application include anticancer drugs, antibiotics, antifungals, antivirals, antidiabetics, NSAIDS, vitamins and minerals.

The humic substances act by improving the dissolution, solubility and permeability characteristics of problematic drugs and hence enhancing their bioavailability. The mechanism for such an increase by humic substances could be a combination of complexation, surface sorption, entrapment in voids and partitioning in to hydrophobic micro-environment of micelles and pseudomicelles and surfactant and/or wetting action of these substances. Various forces are involved when molecules interact with humic substances and the exact nature of these forces that ultimately affect an improvement in bioavailability of hydrophobic molecules needs to be further investigated. The nature of interaction and the kinds of forces and mechanisms involved can only become clear once the structure of fulvic acids has been completely elucidated and understood. Their varying chemistry and subsequent behavior in nature is largely dependent on their natural source and their formation in nature and this makes the job of extricating their structure more difficult.

Various groups of scientists have worked in recent years to demonstrate the utility of humic substances for enhancing the therapeutic activity of modern drugs. The authors and co-workers (Saluja, 2001; Sawnani, 2002; Khanna, 2005; Anwer, 2005; Ahmad, 2006; Tyagi, 2006; Karmarkar, 2007; Vashisht, 2007; Mirza, 2007 and Ahmad, 2008) successfully demonstrated the complexing properties of humic substances extracted from shilajit and utilized these unique characteristics for enhancing the bioavailability of a number of modern drugs belonging to different therapeutic classes. Further research in this area may allow the rational designing of drug formulations and new drug delivery system having the desired biopharmaceutical and therapeutic potential, in the near future.

A number of patents have been filed worldwide which describe the usefulness of these agents for drugs belonging to modern systems of medicine.

Rowland, 1995 in US Patent 5,405,613 describes various vitamin and mineral compositions (Table 12.1) containing shilajit. It has been mentioned that the activity of the vitamins and minerals gets potentiated by addition of small quantity of purified shilajit. The inventor found that the addition of a small amount of purified shilajit to multivitamin/mineral preparation caused an increase in the energy level of the whole formulation and brought it near or at the energy level of shilajit.

WO03035094 describes a number of pharmaceutical, nutritional and cosmetic composition based on the use of purified shilajit/fulvic acid. Studies demonstrated that the use of purified shilajit/fulvic acid compositions prepared according to the method described in the invention resulted in the enhancement of therapeutic activity of the formulations (Tables 12.2 to 12.5).

WO 06078424 describes compositions of polyherbal extracts comprising of *Withania somnifera* and *Mangifera indica* and purified shilajit along with minerals and vitamins for use as anti-viral and/or immune-supporting agents (Table 12.6).

Table 12.1: Vitamin/mineral composition according to US Patent 5,405,613

S.No.	Ingredients	Quantity per tablet
1.	Purified Shilajit	0.4 to 10% w/w
2.	Vitamin A	500 to 10,000 I.U.
3.	Beta Carotene	2,000 to 15,000 I.U.
4.	Vitamin D	50 to 400 I.U.
5.	Vitamin E	30 to 400 I.U.
6.	Vitamin C	75 to 1,000 mg
7.	Vitamin B-1	2 to 80 mg
8.	Vitamin B-2	1 to 80 mg
9.	Niacin	4 to 100 mg
10.	Niacinamide	2 to 100 mg
11.	Pantothenic Acid	10 to 500 mg
12.	Vitamin B-6	3 to 100 mg
13.	Folic Acid	0.002 to 1 mg
14.	Vitamin B-12	3 to 100 meg
15.	Biotin	3 to 80 meg
16.	Calcium	40 to 200 mg
17.	Magnesium	30 to 200 mg
18.	Potassium	10 to 200 mg
19.	Iron	1 to 25 mg
20.	Iodine	0.02 to 0.5 mg
21.	Manganese	0.8 to 6 mg
22.	Zinc	1.5 to 30 mg
23.	Chromium	10 to 80 mg
24.	Selenium	20 to 50 mg

Table 12.2: Antidiabetic composition according to WO 03035094

S.No.	Ingredients	Quantity per tablet (mg)
1.	Glibenclamide	1.00
2.	Purified Shilajit/Fulvic Acid Composition	10.00
3.	Lactose	50.00
4.	Microcrystalline Cellulose	50.00
5.	Croscarmellose Sodium	2.00
6.	Magnesium Stearate	1.00

Table 12.3: Antidiabetic tablet composition according to WO 03035094

S.No.	Ingredients	Quantity per tablet (mg)
1.	Pentazocine	5.00
2.	Purified Shilajit/Fulvic Acid Composition	50.00
3.	Lactose	50.00
4.	Microcrystalline Cellulose	50.00
5.	Croscarmellose Sodium	2.00
6.	Magnesium Stearate	1.00

Table 12.4: Antidiabetic tablet composition according to WO 03035094

S.No.	Ingredients	Quantity per tablet (mg)
1.	Beta Carotene 20% 334 iu/mg	31.25
2.	Ascorbic Acid 97%	500.00
3.	Vitamin E Succinate 1210 iu/g	175.00
4.	Selenomethionine 0.5%	20.00
5.	Zinc Monomethionine, 20%	75.00
6.	Purified Shilajit/Fulvic Acid Composition	20.00 – 200.00
7.	Microcrystalline Cellulose	q.s.
8.	Croscarmellose Sodium	10.00
9.	Silicon Dioxide	10.00
10.	Stearic Acid	20.00
11.	Pharmaceutical Glaze	q.s.

Efficacy studies conducted with the compositions according to the invention demonstrated favorable effect of the administration of the polyherbal composition in patients with HIV infection. The composition resulted in a decreased mean viral load, which was associated with good symptomatic improvement.

US patent application 2003/198695 describes an effective herbo-mineral formulation for prevention and treatment of several forms of anemia caused by hemorrhage, parasitic infestation, nutritional deficiencies, etc. in both children and adults. The formulation according to the invention comprises of purified shilajit containing dimeric and/or oligomeric dibenzo-α-pyrones (DBPs) in synergistic admixture with an

Table 12.5: Skin rejuvenating lotion according to WO 03035094

Skin rejuvenating (O/W) lotion	
Ingredients	**% (w/w)**
Phase A	
Polyglyceryl-3 Methyl Glucose Distearate	3.50
Glyceryl Stearate, PEG-100 Stearate	2.50
Dicapryl Ether	5.00
Coco-Caprylate/Caprate	5.00
Propylene Glycol Dicaprylate/Dicaprate	3.00
Almond Oil	2.00
Cetyl Alcohol	1.50
Purified Shilajit/Fulvic Acid Composition (containing 0.1% retinoic acid)	2.00
Phase B	
Glycerin	3.00
Propylene Glycol	3.00
Allantoin	0.20
Methylparaben	0.15
Water, Deionized	q.s.
Phase C	
Phenoxyethanol and Isopropylparaben and Isobutylparaben and Butylparaben	0.50

extract of the *Emblica officinalis* plant containing gallo-ellagi tannoid and one or more added minerals like iron, chromium, copper, zinc, or manganese, or their mixtures, which form a readily absorbable metal-ion complex with said DBP (Table 12.7).

US patent application 2005/245434 describes a number of pharmaceutical, nutritional and cosmetic formulation based on the use of dibenzo-α-pyrone chromoproteins derived from shilajit (Table 12.8).

WO 05041990 describes a composition containing shilajit or its extract which has the activity of enhancing the metabolic function of the entire body, resulting in an improvement in sexual function and an increase in reproductive function, and thus has effects on nutritional tonic, sexual function improvement, infertility treatment, and the like.

Table 12.6: Vitamin/mineral composition according to WO 06078424

S.No.	Ingredients	Quantity per tablet
1.	Purified Shilajit	0.5 to 30 % w/w
2.	*Withania somnifera*	
3.	*Mangifera indica*	
4.	Vitamin A (Beta Carotene)	45,000 IU
5.	Vitamin B-1 (Thiamin)	25 mg
6.	Inositol Hexanicotinate	50 mg
7.	Vitamin B-6 (Pyridoxine HCL)	25 mg
8.	Vitamin B-12 (Cyanocobalamin)	500 mg
9.	Folic Acid	800 mg
10.	Vitamin C (Magnesium Ascorbate)	150 mg
11.	Vitamin E D-α Tocopheryl (Natural)	400 IU
12.	Copper (Sebacate)	750 mcg
13.	Magnesium (Ascorbate, Taurinate, and Oxide	30 mg
14.	Potassium (Citrate)	10 mg
15.	Selenium (L-Selenomethionine)	200 mcg
16.	Silica (from 400 mg of Horsetail Extract)	10 mg
17.	Coenzyme Q10 (Ubiquinone)	10 mg
18.	L-Carnitine L-Tartrate	50 mg
19.	Hawthorn Berry Extract	40 mg
20.	Grape Seed Extract	10 mg
21.	L-Proline	50 mg
22.	L-Lysine (HCL)	50 mg
23.	N-Acetyl Glucosamine	50 mg
24.	Bromelain (2,000 GDU/g)	120 mg
25.	Taurine (Magnesium Taurinate)	50 mg
26.	Inositol (Hexanicotinate)	10 mg

Table 12.7: Herbo-mineral Iron containing Syrup as per US Appl. 2003/198695

S.No.	Ingredients	Quantity per tablet (mg)
1.	Purified Shilajit	250
2.	*Emblica officinalis*	500
3.	Elemental Iron	100
4.	Excipients	q.s.

Table 12.8: Antidiabetic support tablet according to US Patent Appl. 2005/245434

S.No.	Ingredients	Quantity per tablet (mg)
1.	DCPs	0.10 – 50.00% by weight
2.	Vitamin B-6 (as Pyridoxine HCL)	10 mg
3.	L-Arginine	50 mg
4.	L-Lysine Monohydrochloride	50 mg
5.	Cellulose	q.s.
6.	Magnesium Stearate	q.s.
7.	Gelatin	q.s.

WO03068252 describes a composition containing an extract of Russian mumie, optionally in combination with other pharmaceutically active compounds and pharmaceutically acceptable additives, for improving bone growth and treating osteoporosis (Table 12.9).

Table 12.9: Composition according to WO 03068252 for treating Osteoporosis

S.No.	Ingredients	Quantity per tablet (mg)
1.	Dried Mumie Extract	500.00
2.	Lactose	500.00
3.	Talc	5.00
4.	Magnesium Stearate	1.00

While a number of studies have been done which demonstrate the usefulness of humic substances as components of modern drug delivery systems, a lot needs to be done before the full potential of these fascinating components of nature can be exploited in the pharmaceutical industry. Various factors such as the abundant availability, high aqueous solubility, high stability and nontoxic nature favour their exploitation as carriers for modern drugs.

REFERENCES

Acharya, J.T. (1962). Sushruta Samhita by Dalhanacharya, Sanskrit text with Hindi version, Chaukhamba Surabharati Prakashana, Varanasi, India.

Acharya, S.B., Frotan, M.H., Goel, R.K., Tripathi, S.K., and Das, P.K. (1988). Pharmacological actions of shilajit, *Indian J. Exp. Biol.*, **26**: 775–777.

Ali, M., Sehrawat, I., and Singh, O. (2004). Phytochemical investigation of shilajit. *Indian J. Chem.,* **43B**: 2217–2222.

Agarwal, S.P., Ansari, S.H., and Karmarkar, R. (2008a). Enhancement of the dissolution rate of ketoconazole through a novel complexation with humic acid extracted from shilajit, *Asian J. Chem.*, **20(1)**: 380–388.

Agarwal, S.P., Ansari, S.H., and Karmarkar, R. (2008b). Enhancement of the dissolution rate of ketoconazole through a novel complexation with fulvic acid extracted from shilajit, *Asian J. Chem.*, **20(2)**: 879–888.

Agarwal, S.P., Anwer, M.K., and Aqil, M. (2008c). Complexation of furosemide with fulvic acid extracted from shilajit: A novel approach. *Drug Dev. Ind. Pharm.*, **34**: 506–511.

Agarwal, S.P., Khanna, R., Karmarkar, R., Anwer, M.K., and Khar, R.K. (2008d). Physico-chemical, spectral and thermal characterization of shilajit, a humic substance with medicinal properties. *Asian J. Chem.*, **20(1)**: 209–217.

Agarwal, S.P., Khanna, R., Karmarkar, R., Anwer, M.K., and Khar, R.K. (2007). shilajit: A review. *Phytother. Res.*, **21**: 401–405.

Agarwal, S.P., Khanna, R., Karmarkar, and Khar, R.K. (2006). Improved method for the extraction of fulvic and humic acids from shilajit and their characterization, 13th meeting of International Humic Substances Society, Karlsruhe, Germany.

Ahmad, D. (2006). New formulation of melatonin using humic and fulvic acid as bioenhancers, M. Pharm. Thesis, Jamia Hamdard, New Delhi.

Ahmad, D. (2008). Solubility enhancement of naphthalene through complexation with fulvic and humic acids, M. Pharm. Thesis, Jamia Hamdard, New Delhi.

Ahmed R. Al-Himaidi and Mohammed Umar, Ahmed R. Al-Himaidi and Mohammed Umar (2003). *OnLine Journal of Biological Science* **3(8)**: 681–684.

Aiken, G.R., Mcknight, D.M., and MacCarthy, P. (1985). *Humic substances of soil, sediment and water*, New York: Wiley-Interscience.

Amidon, G.L., Lennernas, H., Shah, V.P., and Crison, J.R. (1995). A theoretical basis for a biopharmaceutic drug classification: The correlation of *in vitro* drug product dissolution and *in vivo* bioavailability, *Pharm. Res.*, **12**: 413–420.

Anwer, M.K. (2005). Novel formulation of furosemide using humic acid and fulvic acid derived from shilajit, M. Pharm. Thesis, Jamia Hamdard, New Delhi.

Anwer, M.K., Agarwal, S.P., Aqil, M., and Khanna, R. (2006). Novel complexes of furosemide with fulvic acid extracted from shilajit: Characterization of the interaction in solid state, 58th Indian pharmaceutical Congress, Mumbai.

Anwer, M.K., Matthias, W., Koch, B., Agarwal, S.P., Ali, A., Sultana, Y., and Khanna, R. (2007). Elemental analyses of fulvic acids of shilajit using ultra high resolution mass spectrometry, 55th ASMS conference on Mass spectrometry, Indianapolis.

Anwer, M.K. (2008). Humic substances as bioenhancer for non-steroidal anti-inflammatory agent, Ph.D. Thesis, Jamia Hamdard, New Delhi.

Assiri, A.M., Khan, Z.A., Al-Ashban, R.M., and Shah, A.H. (2003). Salajeet also called shilajit a natural rejuvenator used in traditional medicine: Literature review and toxicity studies. *J. Saudi Chem. Soc.*, **7(2)**: 199–206.

Arias, M.J., Moyano, J.R., and Gines, J.M. (1997). Investigation of the triameterene-β-cyclodextrin system prepared by co-grinding, *Int. J. Pharm.*, **153**: 181–189.

Avena, M.J., Vermeer, A.W.P., and Koopal, L.K. (1999). Volume and structure of humic acids studied by viscometry, pH and electrolyte concentration effects, *Colloid Surface A,* **151(1-2)**: 213–224.

Baboota, S., Khanna, R., and Agarwal, S.P. (2001). Characteristic features of β-cyclodextrin. The most widely used cyclodextrins, *J. Sci. Pharm.*, **2(3)**: 97–101.

Baboota, S., Khanna, R., Karmarkar, R., and Agarwal, S.P. (2003). cyclodextrins in drug delivery systems in controlled and novel drug delivery, Jain, N.K., *ed.*, New Delhi: CBS Publishers.

Badcock, N.R. (1990). Micro-scale method for itraconazole in plasma by reversed-phase high-performance liquid chromatography., *J. Chromatogr.*, **525(2)**: 478–483.

Baert, L.E.C., Verreck, G., and Thone, D. (1997). Antifungal compositions with improved bioavailability, *PCT Patent Appl. WO9744014.*

Bangaru, R.A., Bansal, Y.K., Rao, A.R.M., and Gandhi, T.P. (2000). Rapid, simple and sensitive high-performance liquid chromatographic method for detection and determination of acyclovir in human plasma and its use in bioavailability studies, *J. Chromatogr. B.*, **739(2)**: 231–237.

Barthe, L., Bessouet, M., Woodley, J.F., and Houin, G. (1998a). The improved everted gut sac: A simple method to study intestinal *p*-glycoprotein, *Int. J. Pharm.*, **173**: 255–258.

Barthe, L., Woodley, J.F., Kenworthy, S., and Houin, G. (1998b). An improved everted gut sac as a simple and accurate technique to measure paracellular transport across the small intestine, *Eur. J. Drug Metab. Pharmacokinet.*, **23(2)**: 313–323.

Basavaiah, K., Prameela, H.C., and Chandrashekar, U. (2003). Simple high-performance liquid chromatographic method for the determination of acyclovir in pharmaceuticals, *Farmaco*, **58**: 1301–1306.

Becirevic-Lacan, M., Filipovic-Grcic, J., Skalko, N., and Jalsenjak, J. (1996). Dissolution characteristics of nifedipine complexes with β-cyclodextrin, *Drug Dev. Ind. Pharm.*, **22**: 1231–1236.

Bhandari, A., Novak, J.T., and Berry, D.F. (1996). Binding of 4-monochlorophenol to soil, *Env. Sci. Tech.*, **30(7)**: 2305–2311.

Bharatrajan, R., Hegde, D., and Nerlekar, N. (2001). Improvement of itraconazole bioavailability, *PCT Patent Appl. WO0197853.*

Bhattacharya, S.K. (1995). Shilajit attenuates streptozotocin induced diabetes mellitus and decrease in pancreatic islet superoxide dismutase activity in rats, *Phytother. Res.,* **9**: 41–44.

Bhattacharya, S.K., Bhattacharya, A. and Chakrabarti, A. (2000). Adaptogenic activity of Siotone, a polyherbal formulation of Ayurvedic rasayanas, Indian J. Exp. Biol., 38:119–128.

Bhattacharya, S.K. and Ghosal, S. (1992). Effect of shilajit on rat brain monoamines, *Phytother. Res.*, **6**: 163–164.

Bhattacharya, S.K, Sen, A.P, and Ghosal, S. (1995). Effect of shilajit on biogenic free radicals, *Phytother. Res.*, **9**: 56–59.

Bhaumik, S., Chattopadhyay, S., and Ghosal, S. (1993). Effect of shilajit on mouse peritoneal macrophages. *Phytother. Res.*, **7**: 425–427.

Bhishagratna, K.K., Susruta Samhita Vol. 2, Chowkhamba Sanskrit Series Office, Varansi, India.

Bietti, G., Dubini, E., Maffione, G., and Vidi, A. (1992). Inclusion compounds of nimesulide with cyclodextrins, *PCI Int. Appl. W0 9117, 774.*

Blume, H.H. and Schug, B.S. (1999). The biopharmaceutics classification system (BCS): Class III drugs – Better candidates for BA/BE waiver? *Eur. J. Pharm. Sc.*, **9(2)**: 117–121.

Boulieu, R., Gallant, C., and Silberstein, N. (1997). Determination of acyclovir in human plasma by high-performance liquid chromatography, *J. Chromatogr. B.*, **693(1)**: 233–236.

Calcagno, A.M., and Siahaan, T.J. (2005). Physiological, biochemical, and chemical barriers to oral drug delivery, *In*: Wang, B., Siahaan, T.J., and Soltero, R., (*eds.*), Drug delivery – Principles and applications, New York: Wiley.

Carreno-Gomez, B. and Duncan, R. (2000). Everted rat intestinal sacs: A new model for the quantitation of *p*-glycoprotein mediated-efflux of anticancer agents, *Anticancer Res.*, **20(5A)**: 3157–3161.

Cauwenbergh, G. (1994). Pharmacokinetics of itraconazole, *Mycoses*, **37**: 27–33.

Chefetz, B., Chen, Y., and Hadar Y. (1998). Purification and characterization of laccase from *Chaetomium thermophilium* and its role in humification, *Appl. Environ. Microbiol.*, **64(9)**: 3175–3179.

Chien, Y., Kim, E.G., and Bleam, W.F. (1997). Paramagnetic relaxation of atrazine solubilized by humic micellar solutions, *Env. Sci. Tech.*, **31(11)**: 3204–3208.

Chilom, G. and Rice, J.A. (2005). Glass transition and crystallite melting in natural organic matter. *Org. Geochem.*, **36(10)**: 1339–1346.

Chopra, R.A., Chopra, I.C., and Handa. K.L. (1958). *In*: *Indigenous Drugs of India*, Calcutta: U.N. Dhar & Sons, pp. 457–461.

Chopra, R.N., Bose. J.P., and Ghosh, N.N. (1926). *Indian J. Med. Res.*, **XIV**: 145–155.

Csabai, K., Cserhati, T., and Szejtli, J. (1993). Interaction of some barbituric acid derivatives with hydroxy propyl β-cyclodextrin, *Int. J. Pharm.*, **91**: 15–20.

Dabur Shilajit, Product literature. (2007). http://www.dabur.com/en/products/ Health_Care/Natural_Cures/shilajit/aphrodisiac.asp, accessed, Dec.

Davies, G. and Ghabbour, E.A. (*eds*) (1999). Understanding humic substances, Royal Society of Chemistry, Cambridge, England.

Davies, G., Ghabbour, E.A., and Khairy, K.A. (1998). Humic substances: Structures, properties and uses, Royal Society of Chemistry, Cambridge, England.

De Beule, K. and Van Gestel, J. (2001). Pharmacology of itraconazole, *Drugs*, **61(Suppl.1)**: 27–37.

De Jalon, E.G., Campanero, M.A., Ygartua, P., and Santoyo S. (2002). HPLC determination of acyclovir in skin layers and percutaneous penetration samples, *J. Liq. Chromatogr. R.T.*, **25(20)**: 3187–3197.

De Miranda, P. and Blum, M.R. (1983). Pharmacokinetics of acyclovir after intravenous and oral administration, *J. Antimicrob. Chemoth.*, **12**: 29–37.

Djedaini, F. and Perly, B. (1991). NMR investigation of the stoichiometries in β-cyclodextrin-steroid inclusion complexes, *J. Pharm. Sci.*, **80**: 1157–1162.

Domeizel, M., Khalil, A. and Prudent, P. (2004). UV spectroscopy: A tool for monitoring humification and for proposing an index of the maturity of compost, *Bioresour. Technol.*, **94**: 177–184.

Dressman, J.B. and Reppas, C. (2000). *In vitro-in vivo* correlations for lipophilic, poorly water-soluble drugs, *Eur. J. Pharm. Sci.*, **11(Suppl.2)**: S73–S80.

Duran-Meras, I., Munoz dela Pena, A., Salinas, F., and Rodriquez-caceres, I. (1994). Spectrofluororimetric determination of Nalidixic acid based on host–guest complexation with γ-cyclodextrin, *Analyst.*, **119**: 1215–1217.

Frolova, L.N., Kiseleva, T.L. (1996). Structure of chemical compounds, methods of analysis and process control :chemical composition of mumijo and methods for determining its authenticity and quality, *Pharm. Chem. J.*, **30(8)**: 543–547.

Frund, R., Ludemann, H.D. (1989). The quantitative analysis of solution and CPMAS C-13 NMR spectra of humic material, *Sci. Tot. Environ.*, **81/82**: 157–168.

Fujioka, Y., Mizuno, N., Morita, E., Motozono, H., Takahashi, K., Yamanaka, Y., and Shinkuma, D. (1991). Effect of age on the gastrointestinal absorption of acyclovir in rats, *J. Pharm. Pharmacol.*, **43(7)**: 465–469.

Gaffney, J.S., Marley, N.A., and Clark, S.B. (1996). *Humic and fulvic acids: Isolation, structure and environmental role,* ACS symposium series, American Chemical Society, Washington, DC.

García, B., Mogollón, J.L., López, L., Rojas, A., and Bifano, C. (1994). Humic and fulvic acid characterization in sediments from a contaminated tropical river. *Chem. Geo.*, **118(1–4)**: 271–287.

Ghabour, E.A., Davies, G. (*eds*) (2000). Humic substances: Versatile components of plants, soil and water, Cambridge: Royal Society of Chemistry.

Ghosal, S. (1989). The facets and facts of shilajit, in research and development of indigenous drugs, Dandiya, P.C. and Vohora, S.B. (*eds.*), Institute of History of Medicine and Medicinal Research, New Delhi, pp. 72–80.

Ghosal, S. (1990). Chemistry of shilajit, an immunomodulatory rasayan, *Pure Appl. Chem.*, **62**: 1285–1288.

Ghosal, S. (1992). Shilalit: Its origin and significance, *Indian J. Indg. Med.*, **9(1&2)**: 1–4.

Ghosal S. (1993). Shialjit: Its origin and vital significance, *In: Traditional medicine, ed.* By B. Mukherjee, Oxford – IBH, New Delhi, pp. 308–319.

Ghosal, S. (1994a). The aroma principles of gomutra and karpurgandha shilajit, *Indian J. Indg. Med.*, **11(1)**: 11–14.

Ghosal. S. (1994b). Shilajit odour: Its origin and chemical character, *J. Indian Chem. Soc.*, **71**: 323–327.

Ghosal, S. (2002). Process for preparing purified shilajit composition from native shilajit, US Patent 6, 440, 436.

Ghosal, S. (2003). Delivery system for pharmaceutical, nutritional and cosmetic ingredients, US Patent 6, 558, 712.

Ghosal S. (2005). Oxygenated dibenzo-alpha-pyrone chromoproteins, US Patent application 2005/0233942.

Ghosal S. (2006). Shilajit in perspective, New Delhi: Narosa Publishing House.

Ghosal, S., Bhattacharya, S.K. (1996). Anti-oxidant defense by native and processed shilajit – A comparative study, *Indian J. Chem.*, **35B**: 127–132.

Ghosal, S., Bhaumik, S., and Chattopadhyay, S. (1995a). Shilajit-induced morphometric and functional changes in mouse peritoneal macrophages, *Phytother. Res.*, **9**: 194–198.

Ghosal, S., Kawanishi, K., and Saiki, K. (1995b). The chemistry of shilajit odour, *Indian J. Chem.*, **34B**: 40–44.

Ghosal, S., Lal, J., Jaiswal, A.K., and Bhattacharya, S.K. (1993). Effects of shilajit and its active constituents on learning and memory in rats, *Phytother. Res.*, **7**: 29–34.

Ghosal, S., Lal, J., Kanth, R., and Kumar, Y. (1993). Similarities in the core structures of shilajit and soil humus, *Soil Biol. Biochem.*, **25(3)**: 377–381.

Ghosal, S., Lal, J., Singh, S.K., Dasgupta, G., Bhaduri, M., Mukhopadhyay, M., and Bhattacharya, S.K. (1989). Mast cell protecting effects of shilajit and its constituents, *Phytother. Res.*, **3**: 249–252.

Ghosal, S., Lata, S., and Kumar, Y. (1995c). Free radicals of shilajit humus, *Indian J. Chem.*, **34B**: 591–595.

Ghosal, S., Lata, S., Kumar, Y., Gaur. B., and Misra, N. (1995d). Interaction of shilajit with biogenic free radicals, *Indian J. Chem.*, **34B**: 596–602.

Ghosal, S., Mukherjee, B., and Bhattacharya, S.K. (1995e). Shilajit – A comparative study of the ancient and the modern scientific findings, *Indian J. Indg. Med.*, **17(1)**: 1–10.

Ghosal, S., Mukhopadhyay, B., and Bhattacharya, S.K. (2001). Shilajit: A rasayan (immunomodulator) of Indian traditional medicine. *In*: Emerging Drugs, Vol. 1, Part 1: Molecular aspects of Asian medicines, Westbury, NY: PJD Publications, pp. 425–444.

Ghosal, S., Reddy, J.P., and Lal, V.K. (1976). Shilajit I: Chemical constituents, *J. Pharm. Sci.*, **65**: 772.

Ghosal, S., Singh, S.K., Goel, R.K., Jaiswal, A.K., and Bhattacharya, S.K. (1991). The need for formulation of shilajit by its isolated active constituents, *Phytother. Res.*, **5**: 211.

Ghosal, S., Singh, S.K., Kumar, Y., Srivastava, R.S., Goel, R.K., Dey, R., and Bhattacharya, S.K. (1988a). Shilajit Part III: Anti-ulcerogenic activity of fulvic acids and 4'-methoxy-6-carbomethoxybiphenyl isolated from shilajit, *Phytother. Res.*, **2**: 187.

Ghosal, S., Singh, S.K., and Srivastava, R.S. (1988b). Shilajit Part II. Biphenyl metabolites from *Trifolium repens*, *J. Chem. Res.*, **196**: 165–166.

Goad, L.J., Berens, R.L., Marr, J.J., Beach, D.H., and Holz, G.G., Jr. (1989). The activity of ketoconazole and other azoles against *Trypanosoma cruzi*: Biochemistry and chemotherapeutic action *in vitro*, *Mol. Biochem. Parasit.*, **32**: 179–189.

Goel, R.K., Banerjee, R.S., and Acharya, S.B. (1990). Anti-ulcerogenic and antiinflamatory studies with shilajit, *J. Ethnopharmacol.*, **29**: 95–103

Grasset, L., and Ambles, A. (1998). Structure of humin and humic acid from an acid soil as revealed by phase transfer catalyzed hydrolysis, *Org. Geochem.*, **29(4)**: 881–891.

Hänninen, K., Knuutinen, J., and Mannila, P., (1993). Chemical characterization of peat fulvic acid fractions, *Chemosphere*, **27(5)**: 747–755.

Hardin, T.C., Graybill, J.R., Fetchick, R., Woestenborghs, R., Rinaldi, M.J., and Kuhn, J.G. (1988). Pharmacokinetics of itraconazole following oral administration to normal volunteers, *Antimicrob. Agents Chemother.*, **32(9)**: 1310–1313.

Higuchi, T., and Connors, K.A. (1965). Phase solubility technique, *Adv. Anal. Chem. Instr.*, **4**: 117–122.

Humex K, Product information. (2008). www.uniqueformulations.co.za/documents/products/humex_k.pdf, accessed, Jan.

Humet-R, Product information. (2008). http://www.humet.com/acatalog/abouthumet.html, accessed, Jan.

Humifulvate, product information. (2008). http://www.humet.com/acatalog/humifulvate.html, accessed, Jan.

Hoshino, T., Uekama, K., and Pitha, J. (1993). Increase in temperature enhances solubility of drugs in aqueous solutions of hydroxy propyl cyclodextrins, *Int. J9. Pharm.*, **98**: 239–242.

Jaiswal, A.K., and Bhattacharya, S.K. (1992). Effect of shilajit on memory, anxiety and brain monoamines in rats, *Ind. J. Pharm.* **24**: 12–17.

Joshi, G.C., Tiwari, K.C., Pande, N.K., and Pande, G. (1994). Bryophyte, the source of the origin of shilajit – A new hypothesis, *B.M.E.B.R*, **15(1-4)**: 106–111.

Jung, C.-R., Schepetkin, I.A., Woo, S.B., Khlebnikov, A.I., and Kwon, B.S. (2002). Osteoblastic differentiation of mesenchymal stem cells by mumie extract, *Drug Dev. Res.* **57**: 122–133.

Jung, J.Y., Yoo, S.D., Lee, S.H., Kim, K.H., Yoon, D.S., and Lee KH. (1999). Enhanced solubility and dissolution rate of itraconazole by a solid dispersion technique, *Int. J. Pharm.*, **187(2)**: 209–218.

Kanikkannan, N., Ramarao, P., and Ghosal, S. (1995). Shilajit-induced potentiation of the hypoglycaemic action of insulin and inhibition of streptozotocin induced diabetes in rats, *Phytother. Res.*, **9**: 478–481.

Karmarkar, R. (2007). New bioavailability enhancers from natural sources. Thesis (Ph.D.), Jamia Hamdard (Hamdard University), New Delhi.

Khanna, R. (2005). Novel bioavailability enhancers from natural sources. Thesis (Ph.D.), Jamia Hamdard (Hamdard University), New Delhi.

Khanna, R., Agarwal, S.P., and Khar, R.K. (2005). A novel complexing agent, Indian Patent Application Number 531/Del/2005.

Khanna, R., Agarwal, S.P., and Khar, R.K. (2005). A novel complexes, Indian Patent Application Number 532/Del/2005.

Khanna, R., Karmarkar, R., Anwer, M.K., Agarwal. S.P., and Khar, R.K. (2006). Physicochemcial and spectral characterization of shilajit, a humic substance with medicinal properties 13th meeting of International Humic Substance Society, Karlsruhe, Germany.

Khanna, R., Witt, M., Anwer, M.K., Agarwal, S.P., and Koch, B.P., (2008). Spectroscopic characterization of fulvic acids extracted from the rock exudate shilajit, *Org. Geochem.*, **39(12)**: 1719–1724.

Kiseleva, T.L., Frolova, L.N., Baratova, L.A., and Yus'kovich, A.K. (1996). HPLC study of fatty-acid components of dry mumijo extract, *Pharm. Chem. J.*, **30(8)**: 421–423.

Klocking, R., Helbig, B. (2005). Medical aspects and applications of humic substances, *In*: Biopolymers for medical and pharmaceutical applications, Steinbuchel, A. and Marchessault, R.H. (*eds.*), Weinheim: Wiley-Vch.

Kong, Y.G., But, P.P.H., Ng, K.H., Cheng, K.F., Cambie, R.C., and Malla, S.B. (1987). Chemical studies on a Nepalese panacea-shilajit, *Int. J. Crude Drug Res.*, **25**: 179–182.

Kumar, R. (1993). Census of mycoflora associated with the Shilajit rocks, *Curr. Sci.*, **65(12)**: 984–985.

Kumar, A., Agarwal, S.P., and Khanna, R. (2003). Modified release bilayered tablet of melatonin using α-cyclodextrin, *Pharmazie*, **58(9)**: 642–644.

Laskin, O.L. (1983). Clinical pharmacokinetics of acyclovir, *Clin. Pharmacokinet.*, **8**: 187–201.

La-tone Gold, Product information (2008). http://www.la-medicca.com/products-ayurvedic_la-tone-gold.html, accessed, Jan.

Legen, I., and Kristi, A. (2001). Comparative permeability of some acyclovir derivatives through native mucus and crude mucin dispersions, *Drug Dev. Ind. Pharm.*, **27(7)**: 669–674.

Livens, F.R. (1991). Chemical reactions of metals with humic material, *Environ. Pollut.*, **70**: 183–208.

Loftsson, T., and Brewster, M.E. (1996). Pharmaceutical applications of β-cyclodextrin I. Drug solubilization and stabilization, *J. Pharm. Sci.*, **85**: 1017–1025.

Loftsson, T., and Fririksdóttir, H. (1998). The effect of water-soluble polymers on the aqueous solubility and complexing abilities of β-cyclodextrin, *Int. J. Pharm.*, **163**: 115–121.

Martin, D.B., and Udupa, N. (1995). β-Cyclodextrin for improved stability, delivery of anti-infective drugs, *Indian J. Pharm. Sci.*, **57**: 203–205.

Martin, D., Srivastava, P.C., Ghosh, D., and Zech, W. (1998). Characteristics of humic substances in cultivated and natural forest soils of Sikkim, *Geoderma*, **84(4)**: 345–362.

McDonnell, R., Holden, N.M., Ward, S.M., Collins, J.F., Farrell, E.P., and Hayes, M.H.B. (2001). Characteristics of humic substances in heathland and foreted peat soil of the Wicklow Montains. *Bio. Enviorn.* **101B(3)**: 187–197.

Menard, F.A., Dedhiya, M.G., and Rhodes, C.T. (1990). Physico-chemical aspects of the complexation of some drugs with cyclodextrins, *Drug Dev. Ind. Pharm.*, **16**: 91–96.

Mikami, Y., Sakamoto, T., and Yazawa, K. (1994). Comparison of *in vitro* antifungal activity of itraconazole and hydroxyitraconazole by colorimetric MTT assay, *Mycoses*, **37**: 27–33.

Mitra, S.K., Sunitha, A., and Kumar, V.V. (1998). Evecare (U-3107) as a uterine tonic: pilot study, *Indian Pract* **51**: 269–272.

Miyake, K., Irie, T., Arima, H., Hirayama, F., Uekama, K., Hirano, M., and Okamaoto, Y. (1999). Characterization of itraconazole/ 2-hydroxypropyl-β-cyclodextrin inclusion complex in aqueous propylene glycol solution, *Int. J. Pharm.*, **179(2)**: 237–245.

Mirza, M.A. (2007). Development and evaluation of inclusion complexes of carbamazepine using natural bioenhancers, M. Pharm. Thesis, Jamia Hamdard, New Delhi.

Mizuma, T., Masubuchi, S., and Awazu, S. (1999). Intestinal absorption of acyclovir β-glucoside: Comparative study with acyclovir, guanosine, and kinetin β-glucoside, *Pharm. Res.*, **16(1)**: 69–73.

Moffat, A.C., Osselton, M.D., and Widdop, B. (2004). Clarke's analysis of drugs and poisons, 3rd edn, London, England: Pharmaceutical Press.

Moyano, J.R., Arias, M.J., Gines, J.M., Pervez, J.I., and Rabasco, A.M. (1997a). Dissolution behaviour of oxazepam in presence of cyclodextrins: Evaluation of oxazepam-dimeb binary system, *Drug Dev. Ind. Pharm.*, **23**: 379–385.

Moyano, J.R., Arias-Blanco, M.J., Gines, J.M., Rabasco, A.M., Perez Martinez, J.I., Nior, M., and Giordano, F. (1997b). NMR investigations of the inclusion complexation of gliclazide with β-cyclodextin, *J. Pharm. Sci.*, **81**: 72–75.

Moore, J.W., and Flanner, H.H. (1996). Mathematical comparison of curves with an emphasis on *in vitro* dissolution profiles, *Pharm. Tech.*, **20(6)**: 64–74.

Nadkarni, K.M. (1976). The Indian *Materia medica*, with ayurvedic, unani and home remedies, Bombay: Bombay Popular Prakashan, pp. 23–32.

National Committee for Clinical Laboratory Standards (NCCLS), (2002a). Reference method for broth dilution antifungal susceptibility testing of yeasts. Approved standard Document M27-A2. National Committee for Clinical Laboratory Standards, Wayne, PA.

National Committee for Clinical Laboratory Standards (NCCLS), (2002). Reference method for broth dilution antifungal susceptibility testing of filamentous fungi. Approved standard M38-A, National Committee for Clinical Laboratory Standards, Wayne, PA.

Nayak, D.C., Varadachari, C., and Ghosh, K. (1990). Influence of organic acidic functional groups of humic substances in complexation with clay minerals, *Soil Sci.*, **149(4)**: 268–271.

Nesseem, D.I. (2001). Formulation and evaluation of itraconazole via liquid crystal for topical delivery system, *J. Pharm. Biomed. Anal.*, **26(3)**: 387–399.

Novotny, F.J., and Rice, J.A. (1995). Characterization of fulvic acid by laser-desorption mass spectrometry, *Environ. Sci. Technol.*, **29**: 2464–2466.

O'Brien, J.J., and Campoli-Richards, D.M. (1989). Acyclovir – An updated review of its antiviral activity, pharmacokinetic properties and therapeutic efficacy, *Drugs*, **37**: 233–309.

Palmieri, G.F., Galli Angeli, D., Giovannucci, G., and Martelli, S. (1997). Inclusion of methoxybutropate in β- and hydroxypropyl β-cyclodextrin: Comparison of preparation methods, *Drug Dev. Ind. Pharm.*, **23**: 27–37.

Park, G.B., Shao, Z., and Mitra, A.K. (1992). Acyclovir permeation enhancement across intestinal and nasal mucosae by bile salt-acylcarnitine mixed micelles, *Pharm. Res.*, **9(10)**: 1262–1267.

Peh, K.K., and Yuen, K.H. (1997). Simple high-performance liquid-chromatographic method for the determination of acyclovir in human plasma using fluorescence detection, *J. Chromatogr. B.*, **693**: 241–244.

Pena-Mendez, E.M., Havel, J., and Patocka, J. (2005). Humic substances – compounds of still unknown structure: Applications in agriculture, industry, environment, and biomedicine, *J. Appl. Biomed.*, **3**: 13–24.

Phillips, P. (1997). Unearthing the evidence, *Chem. Br.*, **33(3)**: 32–34.

Piel, G., Pirotte, B., Deleneuville, I., Never, P., Liances, G., Delarge, J., and Delattre, L. (1997). Study of influence of both cyclodextrins and l-lysine on the aqueous solubility of nimesulide: Isolation and characterization of nimesulide-l-lysine-cyclodextrin complexes, *J. Pharm. Sci.*, **86**: 475–480.

Pierard, G.E., Arrese, J.E., and De Doncker, P. (1995). Antifungal activity of itraconazole and terbinafine in human stratum corneum: A comparative study, *J. Am. Acad. Dermotol.*, **32**: 429–435.

Pietro, M. and Paola, C. (2004). Thermal analysis for the evaluation of the organic matter evolution during municipal solid waste aerobic composting process, *Thermochim Acta*, **413**: 209–214.

Pitha, J., and Hoshino, T. (1992). Effects of ethanol on formation of inclusion complexes of hydroxy propyl cyclodextrins with testosterone or with methyl orange, *Int. J. Pharm.*, **80**: 243–247.

Poirier, J.M., and Cheymol, G. (1998). Optimisation of itraconazole therapy using target drug concentrations, *Clin. Pharmacokinet.*, **35**: 461–473.

Primavie, Product information. (2008). http://www.primavie.com/productinfo.html, accessed, Jan.

Qi, H., Nishihata, T., and Tytting, J.H. (1994). Study of the interaction between β-cyclodextrin and chlorhexidine, *Pharm. Res.*, **11**: 1207–1210.

Rao, P.T., and Gupta, M.L. (1977). Rumalaya in cervical and lumbar spondylosis, *Probe* **3**: 184.

Renone, Product information. (2008). http://www.herbscancure.com/renone.html, accessed, Jan.

Rice, J.A., Tombácz, E., and Malekani, K. (1999). Applications of light and X-ray scattering to characterize the fractal properties of soil organic matter, *Geoderma*, **88(3-4)**: 251–264.

Richards, D.M., Carmine, A.A., Brogden, R.M., Heel, R.C., and Speight, T.M. (1983). Acyclovir: A review of its pharmacodynamic properties and therapeutic efficacy, *Drugs*, **26**: 378–438.

Ruggiero, P., Interesse, F.S., and Sciacovelli, O. (1979). [^{1}H] and [^{13}C] NMR studies on the importance of aromatic structures in fulvic and humic acids. *Geochim. Cosmochim. Acta*, **43(11)**: 1771–1775.

Ruggiero, P., Interesse, F.S., Cassidei, L., and Sciacovelli, O. (1980). ^{1}H NMR spectra of humic and fulvic acids and their peracetic oxidation products. *Geochim. Cosmochim. Acta.* **44(4)**: 603–609.

Saluja, A. (2001). Novel formulations of piroxicam using humic acids derived from shilajit, M. Pharm. Thesis, Jamia Hamdard, New Delhi.

Sanghavi, N.M., Mayekar, R., and Frutiwala, M. (1995). Inclusion complexes of terfenadine-cyclodextrins, *Drug Dev. Ind. Pharm.* **21**: 375–381.

Sanz Sanz, F., and Palacio Hernanz, A. (1988). Randomized comparative trial of three regimens of itraconazole for treatment of vaginal mycoses, *Rev. Infect. Dis.*, **9(Suppl.1)**: S139–S142.

Sawnani, B. (2002). Formulation and evaluation of a dosage form containing nimesulide-humic acid complex, M. Pharm. Thesis, Jamia Hamdard, New Delhi.

Schepetkin, I.A., Khlebnikov, A.I., Ah, S.Y., Woo, S.B., Jeong, C.S., Klubachuk, O.N., and Kwon, B.S. (2003). Characterization and biological activities of humic substances from mumie, *J. Agric. Food Chem.*, **51(18)**: 5245–5254.

Schepetkin, I., Khlebnikov, A., and Kwon, B.S. (2002). Medical drugs from humus matter: Focus on mumie, *Drug Development Research*, **57**: 140–159.

Schnitzer, M. (1972). Chemical, spectroscopic and thermal methods for the classification and characterization of humic substances, Proc. Int. Meet. Humic Substances, Nieuwersluis, Pudoc, Wageningen, pp. 293–307.

Schnitzer, M. (1978). Humic substances: Chemistry and reactions, in soil organic matter, *ed.* by M. Schnitzer and S.U. Khan, New York: Elsevier pp. 1–64.

Shargel, L., and Yu, A.B.C. (*eds*) (1992). Applied biopharmaceutics and pharmacokinetics, 3rd edn, Europe: McGraw-Hill Education.

Sharma, P.V. (1978). *In*: Darvyaguna Vijnan, 4th *eds*, Varanasi, India: Chaukkhamba Sanskrit Sansthan.

Sharma, R.K., and Das, B. (1988). Charak Samhita by Agnivesha, Sanskrit text with English translation, Chaukhamba Sanskrit Series, Varanasi, India.

Shin, H.S., Monsallier, J.M., and Choppin, G.R. (1999). Spectroscopic and chemical characterizations of molecular size fractionated humic acid, *Talanta*, **50(3)**: 641–647.

Shilajit Gold, Product literature. (2007). http://www.dabur.com/en/products Health_Care/Natural_Cures/Shiljit_gold/default.asp, accessed, Dec.

Shilajeet information. (2008). http://www.himalayahealthcare.com/ aboutayurveda/cahs.htm#shilajeet, accessed, Jan.

Simpson, A.J., Kingery, W.L., Hayes, M.H.B., Spraul, M., Humpfer, E., Dvortsak, P., Kerssebaum, R., Godejohann, M., and Hofmann, M. (2002). Molecular structures and associations of humic substances in the terrestrial environment, *Naturwissenschaften*, **89**: 84–88.

Singh, M., and Agarwal, S.P. (2002). Inclusion complexation of Sparfloxacin with bCD, A physico-chemical study, *Die Pharmazie*. **57**: 505–506.

Six, K., Berghmans, H., Leuner, C., Dressman, J., Van Werde, K., Mullens, J., Benoist, L., Thimon, M., Meublat, L., Verreck, G., Peeters, J., Brewster, M., and Van den Mooter, G. (2003). Characterization of solid dispersions of itraconazole and hydroxypropylmethylcellulose prepared by melt extrusion, *Part II., Pharm. Res.*, **20(7)**: 1047–1054.

Solenova Libidoplex, Product information. (2008). www.solanova.com/ hl_**Libidoplex**.htm, accessed, Jan.

Spiteller, M., and Schnitzer, M. (1983). A comparison of the structural characteristics of polymaleic acid and a soil fulvic acid, *J. Soil Sci.*, **34**: 525–537.

Srivatsan, V., Dasgupta, A.K., Kale, P., Datla, R.R., Soni, D., Patel, M., Patel, R., and Mavadhiya, C. (2004). Simultaneous determination of itraconazole and hydroxyitraconazole in human plasma by high-performance liquid chromatography, *J. Chromatogr. A.*, **1031(1–2)**: 307–313.

Stella, V.J., and Rajewski, R.A. (1997). Cyclodextrin: Their future in drug formulation and delivery, *Pharm. Res.*, **14(5)**: 556–567.

Stevens, D.A. (1999). Itraconazole in cyclodextrin solution, *Pharmacother.*, **19**: 603–611.

Stevenson, F.J. (1994). Humus Chemistry: Genesis, composition, reactions, 2nd edn., Wiley Interscience, New York.

Sutton, R., and Sposito, G. (2005). Molecular structure in soil humic substances: The new view, *Environ. Sci. Technol.*, **39**: 9009–9015.

Suzuki, L., Ito, K., Fushimi, C., and Kondo, T. (1993). Application of freezing point depressions to drug interaction studies. A study of 1:1 plus complexes between barbiturates and α-cyclodextrin using freezing point depression method, *Chem. Pharm. Bull.*, **41**: 1444–1449.

Szejtli, J. (1988). Cyclodextrin technology, Davies, J.E.D. *ed*, Dordrecht: Kluwer academic publishers.

Tan, K.H. (2003). Humic matter in soil and the environment. Principles and controversies, New York, NY, USA: Marcel Dekker.

Tewari, V.P., Tewari K.C., and Joshi P. (1973). An interpretation of Ayurvedic findings on shilajit, *J. Res. Ind. Med*, **8**: 53–58.

Thaud, N., Sebille, B., Deratanil, A., and Lilievre, G. (1990). Determination of high performance liquid chromatography of the binding properties of charged β-cyclodextrin derivatives with drugs, *J. Chromatograph.*, **503**: 453–457.

Thomas, A.H. (1986). Suggested mechanism for the antimycotic activity of the polyene antibiotics and the *N*-substituted imidazoles, *J. Antimicrob. Chemoth.*, **17**: 269–279.

Tiwari, L., and Agrawal, D.P., Shilajit. (2005). The traditional panacea: Its properties, http://www.indianscience.org/essays/t_es_agraw_shilajit.shtml, accessed, Oct.

Tiwari, V.P., Tiwari, K.C., and Joshi, P. (1973). An interpretation on Ayurvedic finding on shilajit., *J. Res. Ind. Med.*, **8**: 53–60.

Tolpa Peat Preparation. (2008). http://www.torf.com, accessed, Jan.

Tong, W.O., Lach, J.L., Chin, T.F., and Guillary, J.K. (1991). Structural effects on the binding of amine drugs with the diphenyl functionality to cyclodextrins I.A. microcalorimetric study, *Pharm. Res.*, **8**: 951–955.

Tricot, G., Joosten, E., Bogaerts, M.A., and Vande Pitte, J., Cauwenbergh, G. (1987). Ketoconazole *vs* itraconazole for antifungal prophylaxis in patients with severe granulocytopenia: Preliminary results of two nonrandomized studies, *Rev. Infect. Dis.*, **9(Suppl.1)**: 94–99.

Tschapek, M., and Wasowki, C. (1976). The surface activity of humic acid, *Geochim. Cosmochim. Acta.*, **40**: 1343–1345.

Tripathi, Y.B., Shukla, S. and Chaurasia, S. (1996). Antilipid peroxidative property of shilajit. *Phytother. Res.*, **10**: 269–270.

Tyagi, B. (2007). Novel formulation of paclitaxel using fulvic and humic acids as bioenhancers, M. Pharm. Thesis, Jamia Hamdard, New Delhi.

Valsami, G.N., Koupparis, M.A., and Macheras, P.E. (1992). Complexation studies of cyclodextrins with tricyclic antidepressants using ion selective electrodes, *Pharm. Res.*, **9**: 94–98.

Vashisht, A. (2006). Inclusion complexes of meloxicam with natural bioenhancers, M. Pharm. Thesis, Jamia Hamdard, New Delhi.

Visser, S.A., and Mendel, H. (1971). X-ray diffraction studies on the crystallinity and molecular weight of humic acids. *Soil Bio. Biochem.*, **3(3)**: 259–262.

Wandruszka, R., Haakenson, T.J., and Toerne, K.A. (1998). Evaluation of humic acid fractions by fluorimetry and ^{13}C NMR spectroscopy, Spectrochim. *Acta A Mol. Biomol. Spectrosc.*, **54(5)**: 671–675.

Warnock, D.W., Turner, A., and Burke, J. (1988). Comparison of high performance liquid chromatographic and microbiological methods for determination of itraconazole, *J. Antimicrob. Chemother.*, **21(1)**: 93–100.

Wershaw, R.L., Burcar, P.J., and Goldberg, M.C. (1969). Interaction of pesticides with natural organic material, *Environ. Sci. Tech.*, **3**: 271–273.

Woestenborghs, R., Lorreyne, W., and Heykants, J. (1987). Determination of itraconazole in plasma and animal tissues by high-performance liquid chromatography, *J. Chromatogr.*, **413**: 332–337.

Yoo, S.D., Kang, E., Jun, H., Shin, B.S., Lee, K.C., and Lee, K.H. (2000a). Absorption, first-pass metabolism, and disposition of itraconazole in rats, *Chem. Pharm. Bull.*, **48(6)**: 798–801.

Yoo, S.D., Lee, S.H., Kang, E., Jun, H., Jung, J.Y., Park, J.W., and Lee, K.H. (2000b). Bioavailability of itraconazole in rats and rabbits after administration of tablets containing solid dispersion particles, *Drug Dev. Ind. Pharm.*, **26(1)**: 27–34.

Ziauddin, M., Phansalkar, N., Patki, P., Diwanay, S., and Patwardhan B. (1996). Studies on the immunomodulatory effects of ashwagandha, *J. Ethnopharmacol.*, **50**: 69–76.

SUBJECT INDEX

Symbols

A

B

C

D

E

F

I

J

K

P

Q

R

S

T

V

W